EXPERIMENTAL QUALITY:
A strategic approach to achieve and improve quality

*Dedicated to Frenie, Margaret and
our parents, who instilled the importance of
an interminable quest for knowledge*

Contents

Preface

Improving the quality of products and manufacturing processes at low cost is an economic and technological challenge to the industrial engineers and managers in many organizations. Dr Taguchi's experimental design techniques have proved to be successful in meeting this challenge over the past 15 years. However, research has shown that the application of Taguchi methods in small and medium-sized manufacturing companies is limited due to the lack of statistical knowledge required by engineers. Besides there is a serious communication problem between engineers, managers and statisticians in both the academic and industrial world that impedes the effective use of these advanced statistical techniques for solving manufacturing process quality problems. Many industrial engineers would not know what to do with the results from the statistical analysis and therefore need help from statisticians or consultants to interpret the data and to take the necessary action for improvement of the process under investigation.

This book is written in a way that we hope will enable its use across this broad base, both in the academic world and in the industrial setting. It will be useful to practising industrial engineers, quality engineers and engineering managers from all disciplines. Throughout this book the emphasis will be upon the use of simple statistical techniques and modern graphical tools so that both the engineers and managers can understand the benefits of using such advanced statistical techniques for many practical quality problems. The book delineates a practical and strategic methodology for Taguchi experimental design technique in a methodical manner. The method is explained through real industrial case studies, thus making it easy for the readers to follow the methodology with no prior background in the theory of probability and statistics.

Portsmouth   Jiju Antony
                                                                         Mike Kaye

Acknowledgements

We are deeply indebted to a number of people who, in essence, have made this book what it is today.

First, and foremost, on a personal basis, I would like to express my appreciation to Dr John G. Roche who introduced me to the world of quality management and for providing endless support and guidance during my M.Eng. Sc. course at the University College, Galway, Republic of Ireland.

I thank Dr David C. Robinson and Mr Bernard Higgins for enabling my Doctor of Philosophy programme of study, in the Faculty of Technology, University of Portsmouth. I am also greatly indebted to Professor Anand of the Indian Statistical Institute for many hours of discussion on various issues of Taguchi methods, statistical process control and total quality management.

I would also like to extend my gratitude to Dr Renjit K. Roy of NUTEK Inc., discussions with whom helped the progress of this book in many aspects. I thank all the lead engineers and managers in organizations who sought my advice in carrying out their Taguchi case studies and asked me to facilitate the brainstorming sessions. These exercises provided much of the material, especially in the case-studies section of this book.

We would also like to thank the reviewers for their efforts and for the comments and suggestions that have served to deliver this book to the targetted audience. We very much appreciate the editorial help we received from Ms Deborah Millar and Ms Alison Provan throughout the various stages of the preparation of the manuscript.

We also thank our colleagues—Mr William Ryan, Mr Andreas Frangou, Ms Rose Li, Mr Ebrahim Mazharsolook, Mr Marcus Blosch, Mr Nick Capon, Dr Michael Wood, Mr David Preece and Mr Thomas Kurian—for their friendship and continuous support over the years. Our regrets and apologies go to all those who are not named here who have made contributions directly or indirectly to this text.

The graphical outputs illustrated in the book were prepared using Minitab 11 for Windows. Minitab is a registered trademark of Minitab Ltd.

Finally, I thank all members of my family, in particular my brother Dr Saju Antony, and my parents, for their unbounded love, encouragement, and for everything they have sacrificed for me over the years. A round of applause and a big hug to my wife, Frenie, who reviewed every page of the manuscript with me for typographical and grammatical errors.

Dr Jiju Antony is a research fellow with the Portsmouth Business School, University of Portsmouth. He received a BE in Mechanical Engineering from the University of Kerala, South India, M. Eng. Sc. in quality and reliability engineering from the National University of Ireland and a Ph.D. in quality control for manufacturing from the University of Portsmouth. He has published over 30 refereed papers in the areas of reliability engineering, design of experiments, Taguchi methods, total quality management and statistical process control. His primary research areas include design of experiments and Taguchi methods for improving process quality, robust technology development and problem solving using quality tools and techniques. He is currently developing a computer-based intelligent system for teaching advanced statistical quality improvement techniques (ASQIT) to industrial engineers based on the ideas presented in this book.

Professor Mike Kaye joined the University of Portsmouth after working for 10 years in the steel industry. He is currently Professor of operations and quality management within the Portsmouth Business School. His primary areas of research include simulation, statistical process control, total quality management and strategic quality management. He has published over 40 refereed papers in these areas. Professor Kaye received a B.Sc. in mathematics from the University of Bradford and an M.Sc. in statistics from the University of Southampton.

EXPERIMENTAL QUALITY:
A strategic approach to achieve and improve quality

1 INTRODUCTION

1.1 Introduction

Improving the quality of products and manufacturing processes at low cost represents both an economic and a technological challenge to industrial engineers and managers in many organizations. Dr Genichi Taguchi's experimental design techniques have proved to be successful in meeting this challenge over the past 15 years. Dr Taguchi is a Japanese engineer and quality consultant who has been active in the improvement of various industrial products and processes using statistically designed experiments. However, research has shown that the application of Taguchi methods and other advanced statistical techniques in small and medium-sized manufacturing companies is limited due to the lack of statistical knowledge required by industrial engineers. In addition, there is a serious communication problem between engineers, managers and statisticians, in both the academic and industrial world, which limits the effective use of these advanced statistical techniques in solving manufacturing process quality problems. For any problem solving to take place, it is essential that these groups of people work together and communicate to each other. This book delineates a practical and strategic methodology for Taguchi experimental design in a systematic manner for solving manufacturing process quality problems. The method is explained through real-life case studies in manufacturing, thus making it easier for the engineers and managers to follow the methodology with no prior background in the theory of probability and statistics.

1.2   What is design of experiments (DOE)?

Prior to defining the term DOE, it is important to know the role of experimentation, in general, and in manufacturing companies. According to Webster's dictionary, experiment is defined as [1]: "an operation carried out under controlled conditions in order to discover an unknown effect or establish a hypothesis or to illustrate a known law".

Scientists and engineers perform experiments to increase their understanding of a particular phenomenon. Scientists try to establish the relationship between a cause and an effect beyond any reasonable doubt, and conduct experiments to identify the truth about such a relationship. In contrast, engineers are more concerned with finding the cause of the problem and then search for possible solutions to remove the cause.

In the engineering environment, an experiment is a series of tests or trials which produce quantifiable outcomes. Engineers conduct experiments to improve the quality of products or processes. For example, in a certain injection moulding process, the main concern for engineers was to study the parts shrinkage problem. It was not possible to observe directly what caused the shrinkage. Experience may tell us that several factors such as barrel temperature, injection speed, cure time, type of plastic resin being used, etc. may be responsible for this shrinkage problem. An experiment will help engineers under these circumstances to determine which of these factors affect shrinkage the most. In brief, experiments are generally performed to explore, estimate or confirm. Here exploration refers to data gathering to learn about a process or product characteristic. Estimation refers to determining the effects of several factors on a certain process or product characteristic. Confirmation means verifying the predicted results. For the above example, it is important for the experimenter to know at what level each factor should be kept to minimize the shrinkage. Therefore the experimenter has to design an experiment to determine the best combination of factor levels which minimizes the shrinkage of plastic parts.

Design of experiments (DOE), or statistically designed experiments (SDE), is a scientific approach that allows the experimenter to understand a process and to determine how the input variables (factors) affect the output or quality characteristic. In other words, it is a systematic approach to process optimization.

The statistical design of experiments involves the process of designing and planning the experiment so that appropriate data may be collected, which can then be analysed by statistical methods, resulting in valid and objective conclusions. The statistical design of experiments is an alternative to the traditional, inefficient one-factor-at-a-time experimentation, where only one factor may vary at a time while keeping all other factor levels constant. In a statistically designed experiment, we vary different factors simultaneously at their respective levels to determine the effect of each factor on the output (or response). This book describes experiments for factors at two levels only, i.e. a low level denoted by (1 or −1) and a high level denoted by (2 or +1). For industrial experiments, one may classify factors into two types: qualitative and quantitative. Qualitative factors are those where there is no numerical hierarchy to separate the two levels; for example, good and bad, two brands of chemicals, and so on. Quantitative factors, on the other hand, are those that are measurable and continuous, for example, the thickness of wafer, the diameter of a spindle, the strength of steel, and so on. The response is the quantity we measure in an experiment. When the response relates to some measure of quality, then it is called a quality characteristic.

1.3 Why design of experiments or statistically designed experiments?

Design of experiments is a very powerful approach to product and process development and for improving the yield and stability of an ongoing manufacturing process. Kackar [2] uses a very good example which illustrates one way in which DOE can be used for quality improvement. The Ina Tile Company found that the uneven temperature distribution in a kiln caused a variation in size of the tiles produced. This variation often resulted in tiles of poor quality. The quality control approach would have been to devise methods for controlling temperature distribution and to monitor the process output for product conformity. This approach would have increased both manufacturing and quality costs. However, the company wanted to reduce the tile size variability without increasing costs. Therefore, instead of controlling temperature distribution they tried to find a tile formulation that reduced the effect of uneven temperature distribution on tile size. Through a designed experiment they found such a formulation. Increasing the content of lime in the tile formulation from 1% to 5%, reduced the tile size variation by a factor of 10. They cost-effectively reduced the tile size variation, which resulted in an increase in product quality.

In general, design of experiments can be used to:

- study the effect of various factors on the product or process behaviour;
- understand the relationship between the input variables and the output quality characteristic;
- shorten the product or process development time;
- identify the optimal settings of a process which maximize or minimize the response;
- reduce manufacturing costs;
- improve the reliability of products;
- reduce variability in product functional performance.

1.4   Three approaches to design of experiments— Classical, Taguchi and Shainin

Within the world of design of experiments, there are basically three approaches for improving product and process quality—the classical approach developed by Sir Ronald Fisher in the early 1920s, the Taguchi orthogonal array (OA) designs used by Dr Genichi Taguchi in the late 1940s, and finally Shainin's approach to statistically designed experiments (SDE) in the 1950s.

1.4.1   Classical design of experiments

In the early 1920s, Sir Ronald Fisher used statistical methods for experimental design at the Rothamsted Agricultural Field Research Station in London England. His initial experiments were concerned with determining the effect of various fertilizers on different plots of land. The final condition of the crop was not only dependent on the fertilizer but also on a number of other factors (such as the underlying soil condition, moisture content of the soil, etc.) of each of the respective plots. Fisher used methods of statistical experimental design and analysis, which could differentiate between the fertilizer effects and the effects of other factors. He and his fellow co-worker Frank Yates [3] were the primary contributors to the early literature on experimental design and analysis. Fisher's methods for effective experimentation were a fundamental break from the scientific tradition of varying one factor at a time. These experimental design techniques have only recently been widely accepted in manufacturing companies.

Design of experiments was liberated from its agricultural roots by George E.P. Box in the 1950s. Box and his fellow co-workers introduced the concept of response surface methodology (RSM) and evolutionary operation (EVOP), which were major breakthroughs in the adaptation of experimental design to manufacturing. RSM is often used by researchers in different fields to determine the optimum values for controlled variables (i.e. process variables that can be controlled during the experiment) to optimize (i.e. minimize or maximize) the response. For example, if the response is the efficiency of a process, then maximization of the response is the objective to be achieved by the experimenter. On the other hand, if the response is the impurity level in a certain chemical process, then minimization would be the objective to be achieved by the experimenter. RSM can be applied to the full-scale production or it can be scaled to a laboratory or the pilot plant. When applied to the full-scale production, the method is known as evolutionary operation [4]. EVOP, introduced by George Box, is a technique that can be used to facilitate continous process improvement. It is the continuous optimization of a process. The technique consists of introducing small changes in the control variables under consideration. These control variables should be optimized continuously to keep the response as close as possible to the desired minimum or maximum value. Lucas [5], Myers [6], Box and Draper [7] and Barker [8] have shown the practical applications of these techniques.

Montgomery [9], Sirvanci [10] and Sheaffer [11] have shown the role and use of classical experimental design techniques for various practical industrial applications; examples include:

- reducing part-shrinkage variability in an injection moulding process;
- investigating the factors affecting the variation in the surface roughness of an engine part;
- optimizing and controlling a wire bonding process.

1.4.2 Taguchi methods

In the early 1950s, Dr Genichi Taguchi, "the father of Quality Engineering" introduced the concept of on-line and off-line quality control techniques known as Taguchi methods. On-line quality control techniques are those activities during the actual production or manufacturing, whereas off-line quality control techniques are those activities during the product (or process) design and development phases. His methods of experimental design were introduced into the United States in the early 1980s. In 1980, Taguchi's introduction of the method to several major American companies, including AT&T, Ford and Xerox, resulted in significant quality improvement in product and process design [12].

Taguchi developed both a philosophy and methodology for the process of quality improvement which was heavily dependent on statistical concepts and tools, especially experiments. His method emphasizes the importance of using experimental design in the following four key areas [13]:

- making products and manufacturing processes insensitive to component variation;
- making products and processes insensitive to manufacturing and environmental variations;
- minimizing variation around a target value of the response;
- life testing of products.

Many experts have shown the application of Taguchi methods for solving various industrial problems; examples include:

- reducing the post-extrusion shrinkage of a speedometer cable casing [14];
- determining the critical process parameters affecting the adhesion of surface mount devices to printed circuit boards [15].

Although the statistical techniques proposed by Taguchi continue to be controversial in the statistical community, his engineering ideas about designing quality into products and manufacturing processes, optimizing for robustness of products, and concentrating on variance reduction around target values are still being applied successfully in manufacturing companies [16].

1.4.3 Shainin methods

Dorian Shainin, another expert in the quality field, developed a different approach to experimental design in 1952. Shainin has developed a viable alternative to both the Taguchi and the classical approach developed by Sir Ronald Fisher, by adopting simple but statistically powerful techniques. Some industries in the United States, especially Motorola, are still practising his seven tools applied to design of experiments for reducing process variability, achieving zero defects and high process capability. The Shainin experiments start with multi-vari charts, followed by variable search or full factorials, and end with scatter-plot optimization [17]. We will not elaborate on Shainin's seven tools applied to design of experiments. A thorough explanation of these is given by Bhote in his book, *World Class Quality— Using Design of Experiments To Make it Happen*. According to Bhote, "within the world of design of experiments, the Shainin methods are the simplest, easiest, and the most cost-effective ways to get to the finish line" [18]. But design of experiments practitioners found that Shainin's approach for experimentation and statistical analysis suffers from the same problems associated with the ineffective one-factor-at-a-time experimentation. Logothetis [19] states that Shainin's recommended procedures for experimentation and analysis concentrate only on the "mean response" (rather than on the "mean" as well as on the response "variability') and rely heavily on costly agreements with suppliers for tightening up tolerances as the only way of improving quality. Nevertheless, it could be a useful technique for engineers to identify the few critical factors with their best or worst levels.

The applications of Shainin's experimental design methodology include:

- component search procedure for identifying the significant factorial effects from a large number of factors [20];
- minimizing the number of solder defects for a wave-soldering process using full factorial design [21].

Taguchi not only emancipated experimental designs from process applications, but also introduced the concept of using designed experiments to make products and processes robust, i.e. to make them insensitive to environmental and manufacturing variations. Taguchi has greatly added to the understanding of the role of designed experiments in quality improvement (both product and process), and has provided a structured approach to designing quality into products and processes. This textbook is mainly focused on the experimental design approach recommended by Taguchi.

1.5 Benefits of Taguchi DOE in manufacturing

The potential benefits of Taguchi methods in manufacturing are illustrated by the following 10 real industrial scenarios. The research has shown that factors at two levels are considered as the simplest and most commonly used designs in a manufacturing environment. Experimenters will go for three levels only if non-linearity effects (curvature effects) in the relationship between the response and control factors (i.e. factors which can be controlled during the experiment and normal production conditions) are to be studied. It is always good practice to start with a small experiment and then build knowledge on it (i.e. sequential approach to experimentation) prior to carrying out a large experiment with factors at three levels in the first place, unless it is unavoidable. Each scenario will encompass the objective of the experiment, a brief description, the results obtained and the benefits of the experiment.

Scenario 1

Objective of the experiment.   To determine the critical process parameters affecting adhesion of surface mount devices to Printed Circuit Boards PCBs.

Brief description of the experiment.   An eight-run experiment was selected to study five factors. An appropriate statistical analysis was performed with the aim of identifying the critical (or most significant) factors affecting adhesion of surface mount devices to PCBs. Significant factors were identified and a confirmatory run was undertaken, based on the determined optimal factor settings. The predicted response was compared against the response observed during the confirmatory run.

Results.   The Taguchi experiment has resulted in an improvement in the adhesion strength and thereby reduced maintenance costs and cycle time when compared to the original process.

Benefit of the experiment.   The adhesion strength was improved by 15% and the variation in adhesion strength was reduced by 20%.

Scenario 2

Objective of the experiment.   To determine the optimal operating levels of factors that reduce porosity in a reinforced reaction injection moulding process.

Brief description of the experiment.   An eight-run experiment was chosen to study five factors. Three factors were identified to be statistically significant and have significant effect on porosity. A confirmatory run was carried out to verify the results.

Results. The Taguchi experiment has resulted in a significant improvement in the reaction injection moulding process.

Benefit of the experiment. The average amount of porosity had decreased by 14.88 units (after the experiment) which was a significant improvement. With the forecasted production of 20000 units per year, the expected saving was around £40000.

Scenario 3

Objective of the experiment. To reduce a casting defect called "shrinkage porosity" in a certain casting process.

Brief description of the experiment. Eight experimental runs were used to study seven factors obtained from a brainstorming session. Statistical analysis was carried out with the objective of identifying the most significant factors. Four factors were identified to be statistically significant. A confirmatory run was performed to verify the predicted results.

Results. A substantial reduction in scrap rate has been achieved using Taguchi experimental design methodology.

Benefit of the experiment. The scrap rate dropped from 15% to nearly 4% and therefore the estimated annual saving was nearly £20000.

Scenario 4

Objective of the experiment. To obtain optimum surface roughness on a certain material using a horizontal milling process.

Brief description of the experiment. The experiment was designed to allow the study of seven factors in eight runs. The response was chosen as surface finish measured in micro inches. The analysis has shown that three factors are statistically significant. The optimal parameter settings were determined and a confirmatory run was carried out to verify the predicted results.

Results. Objective of the experiment was met using Taguchi methods.

Benefit of the experiment.   The confirmatory run has shown a substantial improvement in surface finish of about 25%.

Scenario 5

Objective of the experiment.   To determine which factors contribute to excessive noise on constant-voltage transformers.

Brief description of the experiment.   Three factors and three possible interactions were analysed using eight experimental runs. The response was chosen as noise measured in decibels. Statistical analysis was performed to identify the significant factors and their interactions. A confirmatory run was carried out to verify the results.

Results.   The noise level was substantially reduced using Taguchi's experimental design methodology.

Benefit of the experiment.   The rework rate on these transformers has dropped from 25 to 2%, noise output average has dropped in the range of 7–10 decibels, and annual estimated cost savings were around £24 000.

Scenario 6

Objective of the experiment.   To achieve correct hose thickness and to minimize thickness variability in extruded rubber hoses.

Brief description of the experiment.   Seven factors that were causing excess variability problems were selected using brainstorming methods. The response of interest was hose thickness measured in millimetres. Statistical analysis was carried out to determine the final optimal parameter settings. Mean thickness at these factor settings were predicted and a confirmatory run was carried out to verify the results.

Results.   The objective of the experiment was achieved using Taguchi experimental design methodology.

Benefits of the experiment.   The following benefits were achieved using Taguchi's experimental design technique:

- cost savings of more than £10 000;
- reduced scrap rate;
- increased understanding of the process;
- reduced quality costs.

Scenario 7

Objective of the experiment. To improve the yield of an I.C. fabrication process.

Brief description of the experiment. Sixteen experimental runs were performed to study eight factors obtained from brainstorming. The response of interest was epitaxial thickness, measured in micrometres. An analysis based on mean response and response variance was carried out to determine the optimal operational factor settings. A confirmatory experiment was performed in order to verify the results.

Results. The Taguchi experiment has resulted in a significant improvement in the yield of the process.

Benefit of the experiment. The variation in epitaxial thickness dropped by 60% and therefore a significant improvement in the process yield was achieved.

Scenario 8

Objective of the experiment. To optimize and control the wire-bonding process.

Brief description of the experiment. Six factors were studied using 16 experimental runs. Statistical analysis was carried out to determine the optimal factor settings. The response of interest was pull strength, measured in newtons. A confirmatory experiment was performed to verify the results.

Results. Substantial improvement in the average pull strength was achieved using Taguchi's experimental design technique.

Benefit of the experiment. The average pull strength has increased by 30% and therefore customer returns have decreased from 18% to nearly 2%.

Scenario 9

Objective of the experiment. To eliminate customer failures of transformers during "worst case" conditions.

Brief description of the experiment. Five factors and two factor interactions were studied using eight experimental runs. The response chosen was voltage, measured in volts. Analysis was carried out to meet the objective of the experiment. The optimum factor settings were obtained from the analysis and a confirmatory run was carried out to verify the results.

Results. The Taguchi experiment has resulted in a significant improvement by elimination of redesign of transformers.

Benefits of the experiment. The following benefits were gained after the successful completion of the experiment:

- improved customer satisfaction;
- reduced production failures;
- reduced manufacturing and internal reject costs by £41 000 annually;
- improved process capability.

Scenario 10

Objective of the experiment. To minimize the solder defects per million joints of a wave-soldering process using Taguchi methods.

Brief description of the experiment. Five control factors and three noise factors (i.e. factors that cannot be controlled during actual or standard production conditions) were identified using brainstorming. Each factor was kept at two levels as part of an initial investigation of the process. Five control factors were studied using eight experimental runs, and three noise factors were studied using four runs. Statistical analysis was performed to determine the optimal condition.

Results. The optimal factor settings were determined and confirmatory runs were made to verify these optimal settings.

Benefit of the experiment. The number of solder defects were reduced substantially (i.e. by about 15%) after the determination of optimal factor settings for the wave-soldering process.

Scenario 11

Objective of the experiment. To minimize weld leaks on oil-filled railway shock absorbers due to a high rejection rate of more than 25%.

Brief description of the experiment. Eight experimental runs were used to study seven control factors. The data obtained from the experiment were of attribute in nature (i.e. welds either leak or do not leak). The data were represented by 1 (no leak) and 0 (leak). Attribute analysis was performed to determine the optimum condition.

Results. The experiment has identified key variables which can be used to minimize weld leaks on these shock absorbers. No leaking shock absorbers have been produced since the implementation of optimum levels.

Benefit of the experiment. Rejection rates due to leaks were reduced from 30% to about 1%.

Scenario 12

Objective of the experiment. To reduce scrap and rework in a moulding process.

Brief description of the experiment. Eight experimental runs were performed to study four factors. The response was number of good parts (i.e. attribute data). Attribute analysis was performed to accomplish the objective of the experiment.

Results. The objective of the experiment was met by using Taguchi experimental design methodology.

Benefit of the experiment. Annual savings from scrap and rework were estimated to be £30000. The case study of specific applications will be discussed in detail in Chapter 10.

1.6 Problems and gaps in the state of the art

The effective use of experimental design techniques for solving manufacturing process quality problems are limited in manufacturing companies. Some noticeable reasons are:

- The statistical education for industrial engineers at the university level is inadequate. The courses currently available in engineering statistics often tend to concentrate on probability problems and the more mathematically interesting aspects of the subject, rather than the techniques that are more useful in practice.
- There is a serious communication problem between engineers, managers and statisticians, in both the academic and industrial world.
- Lack of expertise and knowledge in engineering in relation to the practically useful statistical techniques.
- Lack of understanding of the benefits of design of experiments to improve the business needs.
- Commercial software products and expert systems in design of experiments or Taguchi methods do not provide any facilities in classifying and analysing the manufacturing process problem, and then selecting the appropriate choice of experimental design for a particular problem.
- Computer software products accentuate the analysis of the data and do not properly address the interpretation of data from Taguchi experiments. Thus many industrial engineers, having performed the statistical analysis, would not know what to do with the results and what to do next.
- Time pressures, in order to produce the product over a short cycle of time, prohibit the application of advanced statistical techniques such as design of experiments and Taguchi methods.

Research has shown that there is an understanding gap in the statistical knowledge required by industrial engineers for solving manufacturing process quality problems using advanced statistical techniques such as design of experiments or Taguchi methods. For example, the lack of statistical knowledge for engineers could lead to problems such as:

- misunderstanding the nature of interactions among factors under consideration for a given experiment;
- misinterpretation of historical data or data from previous experiments;
- inappropriate factor settings for the process; for example, the range of a factor setting is too low or high to observe the effect of that factor on the output (or response).

On the other hand, academic statisticians' lack of engineering knowledge could lead to problems such as:

- lack of measurement system accuracy and precision;
- unreasonably large number of experimental runs;
- inadequate control of nuisance or extraneous factors, which causes excess process variability;
- undesirable selection of process variables (factors) and responses for the experiment.

It is highly desirable for managers to have a basic knowledge in engineering and statistics. Otherwise, it could lead to problems such as:

- low profitability;
- high manufacturing and maintenance costs;
- poor quality and therefore lost competitiveness in the world market place.

Therefore it is very important to note that communication and co-operation among engineers, managers and statisticians plays a vital role in the success of any industrial experimentation for improving manufacturing process quality. The purpose of this book is to bridge that gap by introducing a strategic methodology for Taguchi experiments.

Exercises

1.1 Explain the role of experimentation in an industrial environment.
1.2 Explain the importance of statistically designed experiments in quality improvement of products and processes.
1.3 What are the three approaches to experimental design and how are they different from one another?
1.4 What are the benefits of performing Taguchi-style experiments in manufacturing organizations?
1.5 Describe the problems and limitations in the state-of-the-art methodologies in experimental design.

References

1. Meisel, R.M. (1991) *A Planning Guide for More Successful Experiments*, ASQC Quality Congress Transaction, pp. 174–9.
2. Kackar, R.N. (1985) Off-line quality control, parameter design and the Taguchi method. *Journal of Quality Technology*, **17**(4), 176–87.

3. Fisher, R.A. (1937) *The Design and Analysis of Factorial Experiments*. Imperial Bureau of Social Sciences, London.
4. Aravamutan, R. and Yayin, I. (1993–94) Application of RSM for the maximisation of concora crush resistance of paper board. *Quality Engineering*, **6**(2), 1–19.
5. Lucas, J. (1994) How to achieve a robust process using response surface methodology. *Journal of Quality Technology*, **26**(4), 248–60.
6. Myers, R.H. (1991) Response surface methodology in quality improvement. *Journal of Communications Statistics*, **20**(2), 457–76.
7. Box, G.E.P. and Draper, N.R. (1987) *Empirical Model Building and Response Surfaces*. John Wiley and Sons, New York.
8. Barker, T.B. (1992) *The Evolution of a System for Teaching EVOP*, ASQC Quality Congress Transactions, pp. 302–8.
9. Montgomery, D.C. (1990–91) Using fractional factorial designs for robust process and product development. *Quality Engineering*, **3**(2), 193–205.
10. Sirvanci, M.B. and Durmaz, H. (1993) Variation reduction by the use of designed experiments. *Quality Engineering*, **5**(4), 611–18.
11. Sheaffer, M. (1990) How to optimise and control the wire bonding process. *Solid State Technology*, November, 119–23.
12. Tsui, K.-L. (1992) *An overview of Taguchi method and newly developed statistical methods for robust design*. *IIE Transactions*, **24**(5), 44–57.
13. Box, G.E.P. and Bisgaard, S. (1987) The scientific context of quality improvement. *Quality Progress*, June, 54–61.
14. Quinlan, J. (1985) *Process Improvement by the Application of Taguchi Methods*. Third Symposium on Taguchi Methods, pp. 11–16.
15. Finnis, N. and Morgan, S. (1989) *Taguchi Applications in the Analysis of Surface Mount Device Adhesion*. Second Symposium (European) on Taguchi Methods, December, pp. 193–209.
16. Kacker, R.N. (1986) Taguchi's quality philosophy: analysis and commentry. *Quality Progress*, December, pp. 21–29.
17. Bhote, K.R. (1988) DOE — the high road to quality. *Management Review*, January 27–33.
18. Bhote, K.R. (1990) *A More Cost-Effective Approach to DOE than Taguchi*, ASQC Annual Congress Transactions, pp. 857–62.
19. Logothetis, N. (1990) A perspective on Shanin's approach to experimental design for quality improvement. *Quality and Reliability Engineering International*, **6**, 195–202.
20. Amster, S. and Tsui, K.-L. (1993) Counter examples for the component search procedure. *Quality Engineering*, **5**(4), 545–52.
21. Bhote, K.R. (1988) *World Class Quality — Design of Experiments Made Easier, More Cost Effective than SPC*. American Management Association.

2 THE TAGUCHI APPROACH TO QUALITY IMPROVEMENT

2.1 Taguchi's definition of quality

The term "quality" can be defined in different ways and it has different meanings to different people. However, in order to achieve quality, it is inevitable to have a definition for quality that will reflect the customer's needs and expectations. This book focuses on the definition and concept of quality advocated by Taguchi. Taguchi's approach to quality differs from that of other world-leading quality gurus such as Juran [1], Deming [2], Crosby [3] and Feigenbaum [4]. Taguchi foucuses more on the engineering aspects of quality rather than on management philosophy. He provides a method of achieving robustness (i.e. making insensitive to external variations) for products and processes, which is the key element for any organization to stay competitive in the world market.

Taguchi defines quality as "the loss imparted by any product to society after being shipped to a customer, other than any loss caused by its intrinsic function" [5]. Taguchi believes that the desirability of a product is determined by the social loss it generates from the time it is shipped to the customer; the smaller the social loss the higher the desirability. By "loss" Taguchi refers to the following two categories:

- loss caused by variability of product functional performance;
- loss caused by harmful side-effects.

The variation in a product's functional performance in the consumer's hand would result in a loss not only to the consumer but also to anyone who is affected by the product's performance. For example, if a specific model of car does not start in cold weather, then the car's owner would suffer a financial loss if he had to call a mechanic from the garage. The car's owner is also most likely to be late for work and suffer yet another financial loss. His employer would also suffer a loss as he would lose the services of the employee who is late for work.

An example of a loss caused by a harmful side-effect would be the emission of toxic fumes from the car's exhaust which pollutes the atmosphere.

Although the above are seen as examples of quality losses, the following are not:

- a product that is scrapped or reworked prior to shipment;
- a product resulting in a loss to society through its intrinsic function (for example, cigarette).

These losses Taguchi sees as a cost to the company but not a quality loss. His reasoning is that such situations reflect cultural and legal issues and not engineering issues.

2.2   Understanding variation

This section focuses exclusively on variation arising from industrial experiments. No two units of product by a manufacturing process are ever exactly alike or identical. Each product is different from all others in different ways. For example, the net content of a can of soft drink varies slightly from can to can, the life of a battery is not exactly the same from one to the next of the same brand. The word "variation" in this book refers to **statistical variation** and is defined as the differences among identical units of product in their performance characteristics, such as weight, length, diameter, thickness, life, strength and so on.

The best way to measure variation is to consider multiple measurements under the same experimental conditions. The differences among these measurements generally provide an estimate of the variation. One simple way of summarizing variation in a set of data is to use a histogram.

In quality improvement efforts and problem solving, the collection and

Table 2.1.

Spring Constant (MN/m)						
472	477	482	470	478	470	462
475	466	466	464	476	473	480
472	476	476	468	466	466	472
476	470	478	476	467	470	472
474	480	479	478	466	475	472

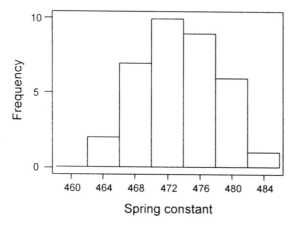

Figure 2.1.   Histogram for the Spring Constant data

use of data play an essential role for sound decisions. The data collected and presented in a tabular format can sometimes be difficult to understand and interpret. Data summary techniques have therefore been developed to help understand the data rapidly and this eases interpretation. One such widely used technique is called a histogram.

The data in Table 2.1 (spring constant values), obtained from a test on 35 glue-laminated beams, were used to construct the histogram illustrated in Figure 2.1. The general shape or distribution of the data enables us to see patterns that are difficult to detect in a sample table of numbers; the patterns, in turn, illustrate the spread (dispersion) of the sample data. Almost all statistical software packages are capable of constructing histograms.

Histograms are easy to construct by hand. The following guidelines will be helpful while constructing a histogram by hand.

- Find the range of the data, which is the difference between the largest and the smallest value in a set of observations.
- Calculate the number of classes (N_c) using equation 2.1 [6]:

$$N_c = 1 + 3.322 \times \log(\Sigma f) \qquad (2.1)$$

where f is the frequency corresponding to each class and Σf is the total frequency. The number of classes should be rounded to the nearest odd number.

- Obtain the class interval by using equation 2.2:

$$\text{Class interval} = \frac{\text{range}}{\text{number of classes} - 1} \qquad (2.2)$$

- Determine the class boundaries (i.e. lower and upper boundaries). The lower boundary is obtained by subtracting half the class interval from the lowest value among all the observations. The upper boundary is obtained by adding the class interval to the lower boundary.
- Count the frequency of data that falls within each of the cells, and draw the histogram.
- To avoid complication and ease interpretation, it is advisable to choose cells of equal sizes.

2.3 Measures of variation

The most common measures of variation are: **the range**, R; **the standard deviation**, SD; **and the mean deviation**, MD.

2.3.1 The range, R

The range of a set of data values is the difference between the maximum value and the minimum value. It is the simplest measure of variation and is generally used when the number of data values is less than 16 [7]. Consider a set of data values: 3, 5, 6, 8, 5, 7, 8, 9, 4, 6. The largest value is 9 and the smallest is 3. Therefore the range is $9 - 3 = 6$.

Although ranges are easy to compute, they can become complicated to interpret and use. This is because ranges tend to increase as the number of data values (or sample size) increases. Therefore one must be concerned about the number of data values when dealing with ranges. Also, the range is highly sensitive to extreme values (in a set of observations, called outliers) as it only uses the two extreme values.

2.3.2 The standard deviation, SD

The standard deviation is a measure of scatter or spread in a set of data values. The standard deviation does not reflect the magnitude of the data values, but only the scatter about the average. The standard deviation provides a more accurate estimate of variation than the range, since it utilizes all the data in the set. Unlike ranges, as the sample sizes increases, the standard deviations do not tend to get larger, but tend to get closer to the true value.

The standard deviation is also sensitive to extreme values, but not as sensitive as the range. The following steps can be applied to calculate the standard deviation:

1. Add the data values and divide this sum by the number of data values. This gives the mean or average.
2. Obtain the deviation of each data value from this calculated average.
3. Square each of the deviations and add the squared deviations. This is called the sum of the squared deviations.
4. Divide the sum of the squared deviations by the number of data values (or sample size), if the true average (i.e. population average) is known. This is called *the **population variance**. Or divide the sum of the squared deviations by the sample size minus 1, if the true average is known and an estimate of the average from data values is substituted instead. This is called *the **sample variance**.
5. Take the square root of the variance. This gives the standard deviation.

The population standard deviation is represented by σ_n, whereas the sample standard deviation is given by s or σ_{n-1}. Mathematically, the population standard deviation, σ_n, is given by:

$$\sigma_n = \sqrt{\frac{\sum_{i=1}^{n}(y_i - \bar{y})^2}{n}} \tag{2.3}$$

where y_i's are observed values, $\bar{y}$ is the average or mean of the population and n is the number of observed values. *The **sample standard deviation** is* obtained by replacing n by $n - 1$. In practice, we nearly always use the sample standard deviation.

If D is the deviation of individual values from the mean $\bar{y}$, then equation 2.3 can be written as:

$$\sigma_n = \sqrt{\frac{\Sigma D^2}{n}} \tag{2.4}$$

Consider a set of data values: 3, 5, 6, 8, 5, 7, 8, 9, 4, 6. The standard deviation (sample) for the given data can be estimated by strictly following the above six steps:

1. Mean or average, $\bar{y} = 61/10 = 6.1$.
2. Deviations, D, of each of the data values from the average are: −3.1, −1.1, −0.1, 1.9, −1.1, 0.9, 1.9, 2.9, −2.1, −0.1.
3. Sum of the squared deviations = $\Sigma D^2 = 32.9$.
4. As we are looking for *the sample standard deviation*, we should divide ΣD^2 by $(n - 1)$ to get *the sample variance*. Therefore the sample variance = 32.9/9 = 3.66.
5. Standard deviation = $\sqrt{3.66} = 1.91$.

2.3.3 The mean deviation, MD

If a set of numbers, say, $y_1, y_2, y_3, \ldots$, and y_n, constituting a sample, has the mean $\bar{y}$, then the differences $y_1 - \bar{y}, y_2 - \bar{y}, y_3 - \bar{y}, \ldots$, and $y_n - \bar{y}$ are called the deviations from the mean. Some of the deviations will be positive and some of them will be negative unless all the y's are same. It is important to note that the sum of the deviations from the mean is always equal to zero.

Since we are interested in the magnitude of the deviations, regardless of their signs (i.e. either positive or negative), we define a measure of variation in terms of the absolute values of the deviations from the mean. If we add the absolute value of deviations of a set of data values from its mean and then divide the result by the number of data values, then we obtain the mean deviation.

The mean deviation, MD, is calculated by the equation:

$$MD = \frac{\Sigma |x_i - \bar{x}|}{n} \tag{2.5}$$

Given the data $\{3, 5, 6, 8, 5, 7, 8, 9, 4, 6\}$, the MD is given by:

$$MD = 15.2/10 = 1.52.$$

This measure has intuitive appeal, but because of the absolute values it leads to serious theoretical difficulties in problems of inference, and it is rarely used [8].

2.4 Variation and its influence on quality

According to Taguchi, variation due to external disturbances (e.g. vibration, humidity, etc.) is the main cause of poor quality. The control and elimination of variation in the product's functional performance should be the main focus of design and process engineers in any organization. Understanding the nature of variation, identifying the causes of variation and then reducing variation using appropriate statistical tools are unduly essential for achieving and improving product and process quality. Before one may begin to reduce process variation, it is important to know "What causes variation?" Throughout this book, the term "variation" refers to statistical variation in manufacturing processes. A process is defined as the combination of inputs, such as materials, machines, manpower, measurement, environment and methods, that results in various outputs which are measures of performance [9]. The total variation in the process may be due to any or a combination of these six sources of variation. In order to obtain the contribution of variation due to each of these sources, one may study the relationship between the inputs and outputs of the process. In other words, we need to study the relationship between the input variables (or factors) and the output (or response) of the process. For example, if we are machining a part, the input variables can be type of machine, type of material, experience of operator, feed rate, speed, type of gauges used, ambient temperature, power supply and so on. The outputs or responses can be thickness, outer diameter, inner diameter or surface finish of the machined part. Under these circumstances, when we have to deal with many variables at the same time and do not know which of these variables are influential on output, Taguchi advocates the use of experimental design techniques. The use of such powerful statistical techniques is to study the potential variables that are likely to be influential on output and to identify those that cause variation in process output. Inconsistency in the product's performance is mainly due to this process variation and therefore the cause of poor quality and low profitability. The use of these experimental design techniques

proposed by Taguchi for reducing variation in processes will be explained
in section 2.9.

2.5 Traditional and Taguchi's approach to quality loss functions

The traditional approach to quality is based on the fact that a company
incurs loss when the performance characteristic of a finished product is not
capable of meeting its specification limits. Performance characteristics are
the final characteristics of a product that determine the product's performance in satisfying users' needs [10]. The picture quality on a TV set is an
example of a performance characteristic. The problem with this approach
is that there is always a need of tight inspection of finished products to
ensure zero defects before the product goes to the hands of the customer.
When the performance characteristic of a product exceeds its specification
limits, then the company incurs massive losses in terms of costs associated
with scrap, rework, warranty or loss of goodwill to the customer. Figure 2.2
illustrates the traditional approach to quality loss function. The loss func-

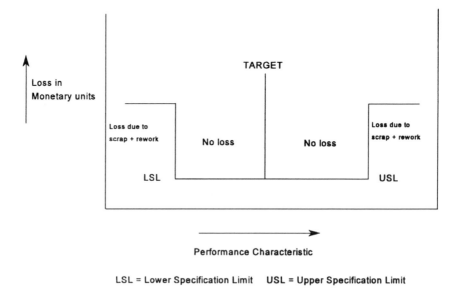

LSL = Lower Specification Limit   USL = Upper Specification Limit

Figure 2.2. Traditional Approach to Quality Loss Function

tion summarizes the relationship between variation in a product's functional performance and costs associated with this variability. The loss function provides a numerical evaluation of product quality [11].

The traditional approach requires more manpower and results in rework and scrap costs that can become prohibitive. It is also unreasonable to assume that a product just inside a specification limit results in no loss to the company [12]. For example, consider the fit between a hole and a shaft. If the diameter of the hole is at the upper limit and the diameter of the shaft is at the lower limit (or vice versa), then the quality of the final fit should be reduced and therefore incurs a loss to the company. However, to develop a quality product at low cost, it is essential that quality must be designed into the product or process at the design stage and focus must be on reducing deviation from the nominal or target value of the performance characteristic of the product. Taguchi suggests a quadratic loss function in analysing these situations. Taguchi's loss function analysis shows that as a product's performance characteristic deviates or moves further away from its target value, an increasing loss will be incurred. The smaller the performance variation about the target value, the better the quality. The degree of customer satisfaction is inversely proportional to the degree of product's performance variation. A product's performance variation is generally determined by several aspects of quality, such as quality of design, quality of conformance and reliability [13]. According to Berton Gunter [14], the concept of the quality loss function is at the heart of Taguchi's quality philosophy. Figure 2.3 illustrates the loss function idea. Taguchi argues that the loss to society is minimum when the product's performance is at the target,

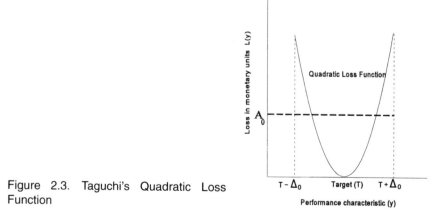

Figure 2.3. Taguchi's Quadratic Loss Function

T and loss increases quadratically as the product's functional performance deviates further from the nominal or target value.

The concept of Taguchi's quadratic loss function as a way to measure quality clearly implies that the goal of engineering design and manufacturing is to produce products and processes that perform "on target with smallest variation". Here quality loss is a measure of deviation from the target value. This quality loss can best be minimized by designing quality into the product right at the design stage. Minimization of quality loss is the only way to stay competitive in today's dynamic global market place.

The monetary loss when a product's functional performance deviates from its target value, T, can be approximated by the equation:

$$L(y) = k(y - T)^2 \qquad (2.6)$$

where: $L(y)$ is the loss in monetary units; k is a constant usually to be determined by cost per unit when the product is outside its specifications (k is called cost constant or quality loss coefficient); y is the performance characteristic (or quality characteristic) of a product; T is the specified target value for the performance *characteristic* of the product; and $[y - T]^2$ is the square of the deviation of performance characteristic from its target T.

The cost constant, k, in the loss function is usually determined by the following equation:

$$k = \frac{A_0}{\Delta_0^2} \qquad (2.7)$$

where Δ_0 is the distance from the target, T, to a tolerance limit (or functional limit), and A_0 is the cost to the customer of repairing, reworking or replacing the product when the functional performance of a product exceeds the tolerance limits (i.e. $T \pm \Delta_0$). These are also called customer's tolerance limits. Equation 2.6 represents the loss function when the performance characteristic or quality characteristic of a product exceeds its tolerance limits and yields the loss to society (customer, manufacturer or anyone else affected by the product).

The following example illustrates the application of a quadratic loss function to a company manufacturing resistors. The target value of the resistance is supposed to be $105\,\Omega$. The tolerance for the resistance is $\pm 2\,\Omega$. If the resistance is out of tolerance, then it costs \$8 to scrap the circuit board.

The cost constant, k (quality loss coefficient), is given by:

$$k = 8/2^2 = 2.$$

Therefore the loss function is given by:

$$L(y) = 2 \, (y - 105)^2.$$

Equation 2.6 represents the expected loss incurred for a single unit when the performance characteristic deviates from its target value. If 'n' units are considered, then the average expected loss is calculated by the following equation:

$$\overline{L}(y) = \left[k \, \Sigma (y_i - T)^2 / n \right] \tag{2.8}$$

where $i = 1$ to n.

$$\overline{L}(y) = k(\text{MSD}) \tag{2.9}$$

where MSD is the mean square deviation. Here MSD is the average of the square of the deviations in the performance characteristic of a product from its nominal value. The above equation can also be expressed as:

$$\overline{L}(y) = k \left[\sigma_y^2 + (\bar{y} - T)^2 \right] \tag{2.10}$$

where: σ_y is the population standard deviation for n values; $\bar{y}$ is the mean or average of performance characteristics of a product; $(\bar{y} - T)^2$ is the square of the deviation in the performance characteristic from the target value; and $(\bar{y} - T)$ is called the bias, which is the deviation of the mean performance from the target value.

Therefore we can write MSD as the sum of the variance and the square of the bias. For practical reasons however, we use s_y^2 instead of σ_y^2. Here s_y^2 is called the sample variance and s_y is the sample standard deviation. The sample standard deviation (s_y) can be obtained easily and is represented by σ_{n-1} in most scientific calculators.

The above loss function can only be applied to certain performance characteristics or quality characteristics of a product, such as dimensions, colour density of a TV set, etc., where one may aim for the target performance. The question is how to minimize the loss incurred due to deviation of the performance characteristic from its target value? This can be achieved effectively by reducing the variability about the target.

Consider two processes X and Y, as shown in Figure 2.4. The figure shows that process X has a much higher proportion of producing parts closer to the target than process Y. Therefore, process X will cause a minimum loss to the society, therefore producing more high-quality products than process Y. For reducing manufacturing process variability, one may have to, first of all, understand the relationship between the response (output) and the process parameters or input variables. Taguchi's experimental design technique can be used in accomplishing this goal. The use of designed exper-

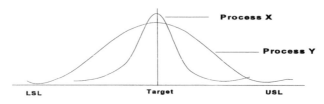

Figure 2.4.   Variation Reduction Around the Target

iments can assist the experimenter in determining which inputs (factors) shift the average or mean of the response, shift the response variability, shift both the average and the variability of the response or have no effect on either the average or the variability of the response.

The quadratic loss function mentioned in equation 2.6 is applicable when the performance characteristic of the product, y, has a finite target value (usually non-zero), and the quality loss function is symmetric on either side of the specified target value.

Example 1. The nominal output voltage for a certain electrical power supply is 115 volts. When the output voltage varies from the nominal by ±10 volts, customers take action at an average cost of $75. What is the value of the cost constant? Write an expression for the quality loss function.

Here the output voltage has a specified target value and therefore the loss function for target performance characteristics can be applied.

Nominal or target output voltage, $T = 115$ volts.

Tolerance or functional limits $\Delta_0 = 10$ volts.

Loss (in $) due to deviation from the target, $A_0 = \$75$.

Therefore the cost constant $(k) = A_0/\Delta_0^2$

$$= \$75/100 \text{ volt}^2 = \$0.75/\text{volt}^2.$$

The loss function is given by: $L(y) = k(y - T)^2$
i.e. $L(y) = 0.75 \ (y - 115)^2.$

Example 2. A manufacturer of magnetic tapes is interested in reducing the variability of the thickness of the coating on the tape. It is estimated that the loss to the consumer is $10 per reel if the thickness exceeds 0.005 ± 0.0004 mm. Each reel has 150 m of tape. A random sample of 15 yielded the following thickness (in millimetres): 0.0047, 0.0048, 0.0049, 0.0053, 0.0053, 0.0051, 0.0048, 0.0049, 0.0047, 0.0053, 0.0052, 0.0054, 0.0046, 0.0051 and 0.0051. Find the average quality loss per reel.

This is a situation in which target is the best-quality characteristic. The loss function for this characteristic is given by: $\bar{L}(y) = k(y - T)^2$.

This is the loss per unit product due to excessive variation of the product's functional performance from its specified target value.

The cost constant is obtained by the equation: $k = A_0/\Delta_0^2$

$$= \$10/(0.0004 \text{ mm})^2$$
$$= \$62\,500000/\text{mm}^2.$$

Therefore the loss function is given by: $L(y) = 62\,500\,000\,(y - 0.005)^2$.

This is the loss per unit product due to excessive variation of thickness from the target.

For a sample of 15 units, the average or expected loss per reel is obtained by: $\bar{L}(y) = 62\,500\,000$ (MSD), where MSD is given by:

$$\text{MSD} = \frac{1}{n} \times \sum_{1}^{n} \{y_i - T\}^2.$$

$\text{MSD} = 1/15\{(0.0047 - 0.005)^2 + (0.0048 - 0.005)^2 + (0.0049 - 0.005)^2 +$
$(0.0053 - 0.005)^2 + (0.0053 - 0.005)^2 + (0.0051 - 0.005)^2 + (0.0048 -$
$0.005)^2 + (0.0049 + 0.005)^2 + (0.0047 - 0.005)^2 + (0.0053 - 0.005)^2 +$
$(0.0052 - 0.005)^2 + (0.0054 - 0.005)^2 + (0.0046 - 0.005)^2 + (0.0051 -$
$0.005)^2 + (0.0051 - 0.005)^2$
$= 6.27 \times 10^{-8}.$

Average quality loss per reel $= 62\,500\,000(6.27 \times 10^{-8})$
$= \$3.92.$

2.6 Determination of manufacturing tolerances

The determination of manufacturing tolerances comprises two steps. The first step is to obtain the societal loss function based on customers' tolerances. Having determined the loss function, the second step is to determine the manufacturing tolerances based on the manufacturer's cost of repairing, adjusting or replacing an item (or product). Let $T \pm \delta$ be the manufacturer's tolerance limits. The manufacturing tolerances are the limits for shipping the product [15]. Let A be the cost incurred by the manufacturer in repairing an item that exceeds the customer's tolerance limits. The idea is that fixing a potential problem prior to shipment will reduce loss in the long run due to wear and tear of the product in its usage. Figure 2.5 illustrates the relationship between the customer's tolerance, $T \pm \Delta_0$, and manufacturer's tolerance, $T \pm \delta$, for a situation in which target is the quality characteristic of interest.

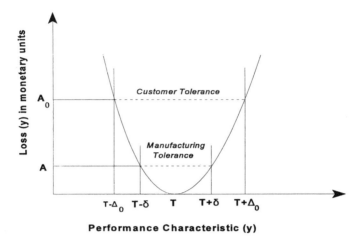

Figure 2.5.   Relationship between the Customer's and Manufacturer's Tolerances

The value of k based on customer tolerance limits is given by:

$$L(y) = k(Y - T)^2$$
$$L(y) = (A_0/\Delta_0^2)(y - T)^2.$$

Now consider the manufacturing tolerance of $T \pm \delta$ and the associated
cost A. Using the loss function, we get:

$$L(y) = (A_0/\Delta_0^2)\left[(T + \delta - T)^2\right]$$

$$\text{or, } A = (A_0/\Delta_0^2)\delta^2$$

$$\therefore \delta = (A/A_0)^{1/2}\Delta_0 \tag{2.11}$$

Because A is usually much smaller than A_0, the manufacturer's tolerance
interval will be narrower than the customer's tolerance interval (refer to
Figure 2.5). As long as the quality characteristic is within δ units from target
value, T, the manufacturer will not spend any extra money. When the quality
characteristic is at δ units from T, the manufacturer may then spend an
amount equal to A, on average, to make an adjustment or repair the
product. It would be wise if the manufacturer spent an amount A as the
customer's loss would otherwise be much more. This customer's loss would
result in an increase of the societal loss.

Example 3. Suppose a garment manufacturer's factory cost of setting and stitching a shirt's collar within some target neck size be \$2.0 (i.e. A) per shirt. Usually this includes an extra measurement and adjustment cost before the garment manufacturer finally stitches the collars. The question is "Where should the shirt manufacturer set his own stitching size tolerance?" This tolerance is the maximum allowable deviation from the specified target collar size to be marked on the shirt.

If it costs the customer \$8 (i.e. A_0) to get the shirt collar that is 1 cm off (i.e. Δ_0) adjusted at a local tailor's shop, then the quality loss coefficient or cost constant is given by:

$$8 = k \cdot (1 \; cm)^2$$

$$\therefore k = 8 \text{ and } L(y) = 8(y - T)^2.$$

This is the loss to society caused by a shirt when its neck size deviates from the target T by 1 cm. This loss function represents the loss imparted to society (i.e. customer, manufacturer or anyone else affected by the product).

The manufacturer's tolerance can now be obtained by using equation 2.11:

$$\delta = (2/8)^{1/2} \times (1 \; cm)$$

i.e., $\delta = 0.5$ cm.

It is important to note that this tolerance is tighter or narrower than the customer's tolerance, because it is cheaper to adjust the collar size while one is manufacturing the shirt than re-stitching by a customer's local tailor. However, manufacturing tolerances need not always be narrower or tighter. The magnitude of manufacturing tolerances should be dependent upon the relative adjustment costs of the customer and the manufacturer.

2.7 Other loss functions

The quadratic loss functions for other quality characteristics—smaller-the-better and larger-the-better characteristics—are discussed below.

2.7.1 Smaller-the-better quality characteristics

In this case, the performance characteristic, y, is continuous and positive with the most desired value of zero (i.e. $T = 0$). Here the loss function, $L(y)$, increases as y increases from zero. Examples of smaller-the-better characteristics include: part shrinkage, tool wear, porosity in casting, leakage current in integrated circuits, pollution from a nuclear power station and so on. The loss function for smaller-the-better quality characteristics is given by:

$$L(y) = k \cdot y^2. \tag{2.12}$$

Because y is continuous and positive, the loss function, $L(y)$, is a one-sided function and therefore cannot accept negative values. The constant, k, can be determined by using equation 2.7. The loss function for this characteristic is shown in Figure 2.6.

Equation 2.12 gives the loss for an individual product. The average loss over many products can be obtained by using equation 2.13:

$$\overline{L}(y) = k \cdot (s^2 + \overline{y}^2) \tag{2.13}$$

where s^2 is the sample variance and $\overline{y}$ is the sample mean of quality characteristic of many products.

Example 4. A certain plastic component manufacturer was suffering from the shrinkage problem for many years. When the shrinkage exceeded 1%, the number of rejected parts increased by 30%. The average cost of replacing the component by a customer is about $15. Determine:

1.   the quality loss function; and
2.   the realistic tolerances for the shrinkage when the rework cost at the end of the production line is $6 per part.

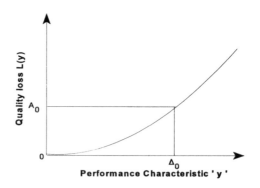

Figure 2.6.   Loss Function for Smaller-the-Better Characteristics

1. This is a situation in which smaller-the-better is the suitable quality characteristic. For these kind of quality characteristics, the loss function is given by:

$$L(y) = k \cdot y^2.$$

The cost constant, k, can be determined by:

$$k = A_0/\Delta_0^2$$
$$\therefore k = \$15/(1\%)^2$$
$$= 15\$\%^{-2}$$
$$\therefore L(y) = 15y^2.$$

2. To determine the realistic production tolerances:

$$L(y) = k \cdot y^2$$
$$\therefore y = [L(y)/k]^{1/2}$$
$$= [6/15]^{1/2}$$
$$= 0.63\%.$$

Therefore, the realistic production tolerance is 0.63% shrinkage.

2.7.2 Larger-the-better quality characteristics

In this case, the performance characteristic is continuous where we would like the characteristic (or response) to be as large as possible. The quality loss, $L(y)$, becomes progressively smaller as the value of the characteristic y increases along the x-axis. The ideal value of this type of characteristic is infinity and the quality loss at that point is zero. Examples of larger-the-better quality characteristics include: the efficiency of engines, the life of batteries, the strength of steel, the yield of a certain chemical process and so on. The loss function for this characteristic is shown in Figure 2.7.

The loss function for larger-the-better quality characteristics is given by:

$$L(y) = k \cdot \left(\frac{1}{y^2}\right) \tag{2.14}$$

where the quality loss coefficient, k, is determined by:

$$k = A_0 \Delta_0^2. \tag{2.15}$$

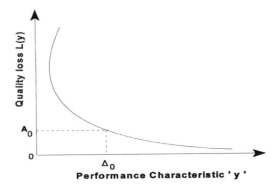

Figure 2.7.   Loss Function for Larger-the-Better
Characteristics

Equation 2.14 yields the loss per unit part or component. The average loss
for many parts or components can be determined by using the following
equation:

$$\overline{L}(y) = k\{[1/\overline{y}^2][1 + (3s^2/\overline{y}^2)]\} \qquad (2.16)$$

where $\overline{y}$ and s^2 are the sample mean and variance respectively. Equation
2.16 can also be written as:

$$\overline{L}(y) = k(\text{MSD})$$

$$\text{where } \text{MSD} = \{[1/\overline{y}^2][1 + (3s^2/\overline{y}^2)]\}.$$

Example 5.   The weld strength of a mildsteel component needs to be
maximized by a certain company. When the weld strength is less than
$0.75\,\text{kg/cm}^2$, some welds crack when subjected to test and cause severe
problems. The average replacement cost of the component to the customer
is about $75. Determine:

1.   the quality loss function; and
2.   the realistic production tolerances for the weld strength, if the
     rework cost at the end of the production line is about $10 per weld.

1. This is a situation in which larger-the-better is the suitable quality
characteristic. For these kind of quality characteristics, the loss function is
given by:

$$L(y) = k \cdot (1/y^2).$$

The cost constant, k, can be determined by:

$$k = A_0 \, \Delta_0^2$$
$$\therefore k = \$75(0.75 \text{ kg}/\text{cm}^2)^2$$
$$= \$42.1 \text{ kg}^2/\text{cm}^4$$
$$\therefore L(y) = 42.19/y^2.$$

2. To determine the realistic production tolerances:

$$L(y) = k \cdot (1/y^2)$$
$$\therefore y = [k/L(y)]^{1/2}$$
$$= [42.19/10]^{1/2}$$
$$= 2.05 \text{ kg}/\text{cm}^2.$$

Therefore, the realistic production tolerance for the weld strength is 2.05 kg/cm^2.

2.8 An industrial application of Taguchi's loss function analysis

Taguchi recommends the use of his loss function analysis to quantify the potential cost savings from the improved optimum product or process design. One may be able to compare the loss associated with n products before and after the experimentation process, from which the savings in monetary units can be estimated.

Engineers involved in an injection-moulding process study wanted to reduce parts shrinkage (i.e. difference between the mould size and part size). A Taguchi-style experiment was performed with the aim of identifying the optimal factor settings that will reduce parts shrinkage. Taguchi's approach to industrial experimentation will be explained in Chapter 5. Data were taken before and after the Taguchi-style experiments. The average unit cost for scrap of the plastic component due to excessive shrinkage (1.25%) was $10. The potential cost savings of the improved process was computed using the Taguchi loss function. The company had a production rate of 45 000 units per year.

Data before the experiment: quality characteristic chosen was shrinkage in millimetres

3.5	2.7	3.2	1.9	1.3	0.8	2.3	4.1	3.4	2.9

This is a situation in which smaller-the-better is the appropriate performance characteristic. The loss function is given by:

$$L(y) = k \cdot y^2$$
$$k = L(y)/y^2$$
$$k = \$10/(1.25\%)^2 = 6.4\$\%^{-2}.$$

Number of observations = 10.

Sum of all observations = 26.1.

Sample mean of observations $(\bar{y})$ = 2.61.

Sample standard deviation (s) = 1.037.

Sample variance (s^2) = 1.075.

For smaller-the-better quality characteristics, the mean square deviation (MSD) for 10 samples is given by:

$$MSD = [s^2 + \bar{y}^2]$$
$$MSD = 1.075 + 6.812 = 7.887.$$

Average loss/unit before the experiment $[\bar{L}_b(y)] = k \cdot MSD = 6.4(7.887) = \$50.48$.

Data after the experiment: quality characteristic chosen was shrinkage in millimetres

1.4	1.1	1.5	1.7	0.7	1.0	1.6	.0.6	1.9	1.2

Number of observations = 10.

Sum of all observations = 12.7.

Sample mean of observations $(\bar{y})$ = 1.27.

Sample standard deviation (s) = 0.427.

Sample variance (s^2) = 0.182.

Mean square deviation (MSD) for 10 samples (after the experiment) is given by:

$$MSD = [s^2 + \bar{y}^2]$$
$$MSD = 0.182 + 1.613 = 1.795.$$

Average loss/unit after the experiment $[\bar{L}_a(y)] = k \cdot \text{MSD} = 6.4(1.795) = \$11.488$.

The average savings per unit can be calculated by substracting the average loss/unit after the experiment from that before the experiment. Therefore, the average savings per unit:

$$\bar{L}_b(y) - \bar{L}_a(y) = \$50.48 - \$11.488$$
$$= \$38.992.$$

With the forecasted production of 45 000 units per year, the annual savings can be determined by multiplying the average savings per unit with the annual production rate:

$$\text{annual savings} = \$38.992 \cdot 45\,000 = \$1\,754\,640.$$

2.9 Taguchi's seven points of achieving quality

The following seven points emphasize the discerning features of Taguchi's approach in achieving and assuring quality.

1. The term "quality" is the deviation of the functional performance of a product from its nominal or target value. The quality of a manufactured product is the total loss incurred by that product to society as soon as the product is shipped to the customer.

    The importance of a particular aspect of quality changes with the nature of the product and the expectations and requirements of the customer. Therefore the specific meaning of the word "quality" changes with the context in which it is being used. Taguchi has defined quality as a loss imparted to society when a product's characteristics do not meet the customer's expectations in terms of tolerances, reliability, serviceability, availability, etc. for which it was designed. The losses imparted by a certain product to the society include:

    • the cost of customer dissatisfaction which may lead to a loss of reputation and goodwill for the company;
    • failure to meet the ideal performance and harmful side-effects caused by the product;
    • loss due to market share and the increasing marketing efforts needed to overcome lack of competitiveness.

2. In a competitive economy, continuous quality improvement (CQI) and cost reduction are necessary for staying in business.

In the present competitive world market place, a business that does not earn a reasonable profit cannot survive for long. The profit is determined by the number of units sold (or market share) and the manufacturing cost. The market share can be increased by providing high-quality products to customers at low costs. Companies determined to stay in business use high quality and low costs as their strategic weapon. Therefore, in a competitive world market, it is inevitable to improve quality and reduce manufacturing costs continuously, for staying in business.

3. A continuous quality improvement (CQI) programme includes continuous reduction in the variation of the product's functional performance characteristics about their specified nominal values or target values.

   The objective of a CQI programme is to reduce the variation of the product functional performance characteristics about their target values. Here "performance characteristics" of a product refers to the quality characteristics that determine the product's performance in meeting and satisfying customer's needs and expectations. The sound clarity on a radio is an example of a performance characteristic. The ideal value of a performance characteristic is called the target value [16]. A high-quality product performs its intended function for a specified period of time under all different operating or environmental conditions. For example, an automatic watch with an intended life span of 7 years has poor quality if it stops or shows incorrect timings after 2 years' use. This performance variation (or inconsistency of performance) is one of the key elements for achieving high-quality products; the smaller the performance variation about the nominal value, the better is the quality.

4. The customer's loss due to variation or deviation in a product's functional performance is often approximately proportional to the square of the deviation of the performance characteristic from its nominal value.

   The concept of Taguchi's quadratic loss function accentuates the importance of continuously reducing performance variation. This has been explained in detail in Section 2.6.

5. The final quality and cost (R&D, manufacturing and operating) of a manufactured product are determined to a large extent by the engineering designs of the product and its manufacturing process.

   A product's field performance is affected by environmental variable (such as ambient temperature, relative humidity, dust and so on), product deterioration and manufacturing imperfections. Countermeasures against variation in the functional performance caused by environmental variables and product deterioration can be built into the product only at the product design stage. The manufacturing imperfections in a product are determined largely by the design of the manufacturing process. Shewhart emphasized the importance of bringing a process to a state of statistical control in improving the design of an existing process.

6. Variation in product or process performance can be reduced by exploiting the non-linear effects of the product or process parameters on the performance characteristics.

The desired target value of the performance characteristic of a product or process depends on how robust (i.e. insensitive) is the process or product against external disturbances such as relative humidity, ambient temperature, etc., which are difficult to control during normal production conditions. The robust design (sometimes called parameter design) is an efficient and effective way of achieving robustness for products and processes by exploiting the non-linearity of factor effects. In other words, by exploiting non-linearity, one can reduce the quality loss without increasing the product cost [17].

7. Statistically planned and designed experiments can identify efficiently and reliably the settings of product or process parameters that reduce performance variation.

The objectives of statistically planned and designed experiments are:

• to identify the settings of design parameters (or control factors) at which the effect of noise factors on the performance characteristic is minimum;

• to identify the settings of design parameters that reduce cost without hurting quality;

• to identify the settings of design parameters that have a large influence on the mean response;

• to identify the settings of design parameters that have a significant influence on the response variability or variability in product's performance characteristics.

2.10 Taguchi's quality engineering system

Dr Taguchi, one of the leading quality gurus of the world, has made a significant contribution to "quality engineering", which deals with making high-quality products at low cost. Taguchi accomplished this objective by utilizing statistical experimental design techniques. The ultimate purpose of quality engineering is to reduce variability in the product's functional performance, and thereby reducing quality loss. It is important to note that the philosophies of Taguchi's quality engineering and classical design of experiments are different [18]. In classical design of experiments, the emphasis is statistical in nature, while in quality engineering, the emphasis is engineering in nature. Moreover, in classical design of experiments, the objective is to study the factors and interactions that explain the product or process behaviour using statistical techniques. In contrast, optimizing the performance of a product or process is the goal in Taguchi's quality engineering philosophy.

The purpose of the development of Taguchi's quality engineering system is to produce products insensitive to undesirable and uncontrollable variations. In other words, making products robust is the goal of Taguchi's quality engineering system. In order to achieve robustness, quality control efforts must begin in the product and process design and development stages, and be continued through production or manufacturing engineering and production operation phases [19].

Taguchi classifies the quality engineering system into two categories:

1. on-line quality control system;
2. off-line quality control system.

2.10.1   On-line quality control system

Here Taguchi refers to quality control activities during actual production or manufacturing phase as "on-line quality control". Some examples of on-line quality control methods include: control charts, process capability studies and so on. Taguchi does not recommend the conventional control chart methodology during normal production, which often attempts to find the cause of variability and remove it [20]. He strongly believes that the main objective of an on-line quality control system should be prevention and not rectification. The variability in the production process should be diagnosed and dealt with before it becomes a cause for concern.

Generally, Taguchi's on-line quality control methods can be used to remove the following sources of variability during normal production:

*   variability in raw materials and purchased components or devices;
*   process trend or drift;

- tool wear;
- machine ageing;
- variability in execution due to human error, and so on.

Taguchi divides the on-line quality control system into three stages:

Process diagnosis and adjustment/preventive maintenance. In this stage, the process will be diagnosed at regular intervals. During the diagnosis period, a product is inspected. If the unit is not faulty, production is continued. If it is faulty, production is stopped and an attempt must be made to discover what causes the problem. The proper process adjustment can be made at this stage and production can be restarted with the adjusted process. Another unit is then produced under the adjusted process. If the unit is good, then normal production restarts. Alternatively, preventive adjustments can be made when imminent failure is diagnosed.

Prediction and correction. This is also called feedback and feedforward control. A quantitative characteristic to be controlled is measured at regular intervals, and the measured value is then used to predict the mean characteristic value of the product on the assumption that the production is continued without adjustment. If the predicted value differs from the target value, then the level of a "signal-factor" is modified to reduce the difference. A signal factor is one that affects only the mean characteristic of a product and not the variability.

Measurement and action. Taguchi refers to this stage as "inspection" [21]. Here, each unit produced is measured, and if it is out of specification, it is reworked or scrapped. This method of quality control deals only with the product, while the stages described above deal mainly with the process. The authors will not elaborate on Taguchi's on-line quality control methods further here. The interested reader should consult some quality control books written by Taguchi for details [21].

Recent research has shown that Taguchi's on-line quality control methods strongly aid in attaining continuous quality improvement which will extend an impetus towards sustained total quality management (TQM) [22]. A major survey, which ranged from available literature in international journals and proceedings of national and international conferences to the interviewing of manufacturers, revealed that the industrial world needs to incorporate the benefits of Taguchi's on-line quality control methods as a key criterion to effect a fully fledged TQM programme.

2.10.2   Off-line quality control system

Taguchi refers to quality control activities at the product planning, design and production engineering phases as "off-line quality control". Off-line quality control methods are used to improve product quality and manu-facturability; reduce product development, manufacturing and lifetime costs; and improve product reliability. Effective quality control requires off-line quality control activities that focus on quality improvement and not quality evaluation. Some examples of off-line quality control activities include product prototype tests, accelerated life tests, reliability tests and so on. This book will focus on off-line quality control activities for improving the product and process quality in manufacturing organizations. The manufacturing fraternity today views the application of Taguchi's off-line quality control methods as a necessary step towards the implementation of a prevention-based quality system, which is a prerequisite for successful total quality management.

Taguchi has developed a structured and systematic off-line quality control methodology for optimizing both product and manufacturing process design. The methodology is unique in that its primary goal is not simply to reduce variability and thereby improve quality; rather, the goal is to improve product and process quality while reducing cost. Taguchi identifies three stages in the design and development of a product or manufacturing process: system design, parameter design and tolerance design.

Figure 2.8 illustrates the three-stage approach for improving quality within Taguchi's off-line quality control/engineering system. The product design stage is concerned with identifying customer needs and expectations, designing a product to meet those needs, while ensuring that the product can be manufactured consistently and economically. The process design stage is concerned with determining the machines and equipment required, the process specifications and standards and the process effectiveness [23].

System design (SD).   This is the first phase of Taguchi's design strategy which requires the technical knowledge and experience of engineers and scientists to visualize the creation of a product or process that provides the function the customer is expecting. System design requires an understanding of both the customers and the manufacturing environment. At this stage, innovation, technology and creativity are the crucial engineering factors, and design engineers are looking for the best available technology to achieve the desired performance from the product [24]. During the product design and development stage, system design comprises the development of a functional prototype design which defines the product design charac-

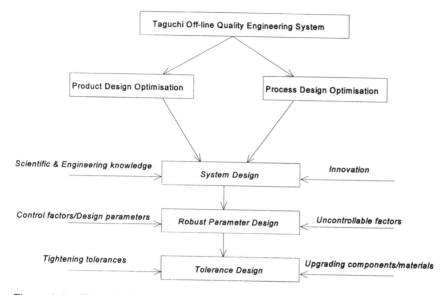

Figure 2.8. Taguchi Off-line Quality Engineering System

teristics or parameters and determination or selection of parts, materials, components and the system assembly. In contrast, during the process design stage, system design comprises the determination of the suitable manufacturing process that can produce the product within the specified tolerances at the lowest cost.

Robust parameter design (RPD). The objective of parameter design (during the product design and development stage) is to identify the settings of the product design characteristics or parameters that will make the product's functional performance insensitive to undesirable sources of variation, i.e. environmental variations, product deterioration and manufacturing variation. In other words, the goal of parameter design is to determine the best product parameter values, which will render "robust function". During manufacturing, the goal of parameter design is to identify the process parameter (or process variable) settings that will minimize the effect of manufacturing variations on the process performance and effectiveness. Typical examples of manufacturing variations include temperature variation, raw material variation, tool-condition variation, operator variation and so on. These variations, as well as several unidentified (or hidden) factors, can cause non-uniformity in the production processes, resulting in either out-of-specification product or non-uniformity of process output.

Taguchi advocates the use of experimental design as a powerful technique to identify the effect of different process variables on the output performance. The objective of experimentation during this stage is to produce a product that has minimal variation in its quality characteristics—the features that matter to customers [25]. Taguchi claims that this is the most important stage of any product or process design, where the biggest gains are made at the lowest cost. Until recently, many US companies have gone straight to the tolerance design after employing the system design, which often leads to expensive products. The Japanese concentrate on the intermediate stage of parameter design and it is at this stage that quality of products and processes can be improved substantially at low cost. A recent survey has shown that US engineers focus 70% of their efforts on system design (compared to Japanese engineers' 40%) and only 2% on parameter design. Japanese engineers, in contrast, apply 40% of their efforts to parameter design, the stage where costs can be controlled most efficiently [26].

Tolerance design (TD).   Tolerance design (TD) is performed after the completion of a parameter design. This stage applies during the product design or process design if the reduction in variation achieved from the parameter design is deficient. During this stage, engineers may tighten the tolerances around the nominal values of the product or process design parameters obtained from the parameter design to achieve the desired variability. It also involves upgrading of materials and components which would have a significant effect on improving quality, while a substantial increase in costs is also estimated. TD determines the trade-off between the quality loss and cost. For example, the narrower the tolerance band, the more costly it becomes to manufacture the product. On the other hand, the wider the tolerance hand, the larger the quality loss and therefore the greater the risk of product non-conformity. It is good practice to select the tolerance design once critical parameters influencing the product output performance are selected during the parameter design. It is estimated that US engineers focus 28% of their efforts on tolerance design, compared to Japanese engineers whose focus is 20%.

While both parameter design and tolerance design are considered optimization phases, the attitudes are substantially different. In parameter design, the strategy is to make products and processes robust against undesirable sources of variation, without increasing costs. The strategy in tolerance design is to remove the sources of variability by increasing costs. Tolerance design involves evaluating the quality versus cost trade-off, and requires spending money to assure quality.

Exercises

2.1 What do you understand by the term "variation" and what are its measures?

2.2 Distinguish between the traditional quality loss function and Taguchi's quadratic loss function analysis.

2.3 Compare Taguchi's loss functions for target-the-best, smaller-the-better and larger-the-better quality characteristics.

2.4 Explain why customer and producer loss are both related to loss to society.

2.5 A certain product has a quality characteristic with specifications 20 ± 2. The product becomes defective if the characteristic exceeds the specification limits. The cost to the consumer for getting it fixed (i.e. repair or adjustment) is $10. Suppose the manufacturer can rework the product, prior to shipping the product to the consumer, at a cost of $4 per product. What should be the manufacturer's specifications?

2.6 The customer tolerances for the height of a steering mechanism are 1.5 ± 0.020 m. For a product that just exceeds these limits, the cost to the consumer for getting it repaired is $50. Ten products were randomly selected and yielded the following heights in metres: 1.53, 1.49, 1.50, 1.49, 1.48, 1.52, 1.54, 1.53, 1.51 and 1.52. Calculate the average loss per unit of the product.

2.7 Distinguish between off-line and on-line quality control.

References

1. Juran, J.M. (1989) Juran on Leadership for Quality, New York: Free Press.
2. Deming, W.E. (1986) "Out of the Crisis", Cambridge, MA: MIT Center for Advanced Engineering Studies.
3. Crosby, P.B. (1979) "Quality is Free", New York, McGraw-Hill Publishers.
4. Feigenbaum, A.V. (1983) "Total Quality Control", Third edition, New York, McGraw-Hill Publishers.
5. Taguchi, G. (1986) "Introduction to Quality Engineering—Designing Quality into Products and Processes", Kraus International, Asian Productivity Organisation, Japan.

6. Kennedy, J.B. and Neville, A.M. (1986) "Basic Statistical Methods for Engineers and Scientists", Harper & Row Publishers.
7. Wheeler, D.J. (1988) "Understanding Industrial Experimentation", SPC inc.
8. Freund, John E. (1988) "Modern Elementary Statistics", Prentice-Hall.
9. Kiemele, M.J. and Schmidt, S.R. (1992) "Basic Statistics—Tools for Continuous Improvement", Air Academy Press.
10. Kackar, R. (1985) "Off-line Quality Control, Parameter Design and the Taguchi Method", Journal of Quality Technology, Vol. 17, No. 4, pp. 176–188, October.
11. Chan, L.K., Cheng, S.W., and Wang, Z.M. (1991) "Applications of Loss Function and Tolerance Design", ASQC Quality Congress Transactions, pp. 539–546.
12. Schmidt, S.R. and Launsby, R.G. (1992) "Understanding Industrial Designed Experiments", Air Academy Press.
13. Juran, J.M. (1979) "Quality Control Handbook", 3rd edition, McGraw-Hill Publications.
14. Gunter, B. (1987) "A Perspective on the Taguchi Methods", Quality Progress, June.
15. Mitra, A. (1993) "Fundamentals of Quality Control and Improvement", Macmillan Publishers.
16. Kackar, R.N. (1986) "Taguchi's Quality Philosophy: Analysis and Commentry", Quality Progress, pp. 21–29, December.
17. Phadke, M.S. (1989) "Quality Engineering Using Robust Design", Prentice-Hall International.
18. Wilkins, J. (1991) "Introduction to Parameter Design", Tutorial, 9th Symposium on Taguchi Methods, pp. A-1–A19.
19. Taguchi. Genichi, Elsayed, E.A., Hsiang, T. (1989) "Quality Engineering in Production Systems", McGraw-Hill Publishers.
20. Logothetis, N. and Wynn, H.P. (1989) "Quality Through Design–Experimental Design, Off-line Quality Control and Taguchi's Contributions", Oxford Science Publication, Oxford.
21. Taguchi, G., "Introduction to Quality Engineering", Asian Productivity Organisation, 1986.
22. Aravindan, P et al. (1995) "Continuous Quality Improvement through Taguchi's On-line Quality Control Methods", International Journal of Operations and Production Management, Vol. 15, pp. 60–77.
23. Antony, J. (1995) "An Overview of Taguchi Off-line Quality Engineering System", Journal of Quality Quest, Vol. 4, No. 4, October, pp. 14–18.
24. Antony, J., Kaye, M. and Robinson, D.C. (1996) "Sorting Out Problems", Manufacturing Engineer", IEE, October, pp. 221–223.
25. Barnard, S.C. and Brown, A.D. (1989) "Why it is Good to be Noisy", Second European Symposium on Taguchi Methods, pp. 1–18.
26. Kumar, S. and Tobin, M. (1990) "Design of Experiment is the Best way to Optimise a Process at Minimal Cost", IEEE Symposium, pp. 166–173.

3 THE TAGUCHI APPROACH TO INDUSTRIAL EXPERIMENTATION

3.1 Traditional approach to experimentation

In the traditional approach to experimentation, experimenters may vary one factor or variable, keeping all other factors in the experiment fixed. This is also called the one-factor-at-a-time approach to experimentation. Here a factor refers to a controlled or uncontrolled variable where influence upon a response (or output) is studied during the experiment [1]. A factor can be either qualitative (i.e. different detergents, machines, vendors, catalysts and so on) or quantitative (i.e. pressure, time, temperature, speed and so on).

Consider three factors (say, A, B and C), each kept at two levels; level 1 and level 2 respectively (see Table 3.1). Here a "level" is a specified value or setting of the factor being examined in the experiment. For example, if the experiment is to be performed at three different speeds, then we can say that the factor, speed, has three levels.

Here an experimental trial (or run) shows the combination of levels for all factors to be tested for an experiment. For example, in Table 3.1, factor A (level 1), factor B (level 1) and factor C (level 1) would constitute the first experimental trial or run. In the first experimental trial, it is obvious

Table 3.1. Version of the "One-Factor-at-a-Time method"

Experimental trial or run	A	B	C	Results
1	1	1	1	y_1
2	2	1	1	y_2
3	1	2	1	y_3
4	1	1	2	y_4

that we keep all factors at level 1. In the second trial, only the level of factor A has changed to level 2, keeping the levels of other two factors at level 1. The difference in the results (i.e. observed output values: y_2—y_1) between these two experimental trials in Table 3.1 provides an estimate of the effect of factor A. An "effect" here refers to the change in output (e.g. efficiency, yield, thickness, life, strength and so on) which we measure during the experiment due to the change in factor levels (i.e. from level 1 to level 2). If there is more than one observation (or result), then the difference in the average response or output value will provide an estimate of factor effect. The effect of factor A has been estimated when both factors B and C were at level 1 and therefore there is no guarantee whatsoever that factor A will have the same effect when the conditions of other factors change. Therefore the one-factor-at-a-time approach to experimentation can be misleading and often lead to wrong and unsatisfactory conclusions. In today's dynamic industrial environment, we need a reliable estimate for factors under study. In other words, the estimate of factor effect on output should be consistent and reproducible when the conditions of other factors change.

Scenario

Suppose an experimenter is interested in determining the yield of a certain chemical process at two temperatures, say T_0 and T_1, and at two pressures, P_0 and P_1, respectively. Assume that the experimenter has used one-factor-at-a-time approach to determine the optimum condition. Here, optimum condition refers to the combination of levels of temperature and pressure where the maximum yield is obtained.

The first step of the experimenter was to keep the temperature constant (T_0) and then vary the pressure from P_0 to P_1. The experiment was repeated twice and the average yield was calculated. The results are shown in Table 3.2.

Table 3.2.   Version 1 of the One-Factor-at-a-Time
Approach

Temperature	Pressure	Average Yield (%)
T_0	P_0	51
T_0	P_1	61

Table 3.3.   Version 2 of the One-Factor-at-a-Time
Approach

Pressure	Temperature	Average Yield (%)
P_0	T_0	51
P_0	T_1	58

The next step was to vary the temperature from T_0 to T_1, keeping the pressure constant (P_0). The results are recorded as shown in Table 3.3.

Here the experimenter has obtained the average yield values corresponding to only three combinations of temperature and pressure: (T_0, P_0), (T_0, P_1) and (T_1, P_0). The experimenter concluded from the above experiment that the maximum yield of chemical process will be attained corresponding to (T_0, P_1). Therefore the optimum condition obtained by the experimenter was: temperature at level 1 (i.e. T_0) and pressure at level 2 (i.e. P_1).

The question then arises as to what should be the average yield corresponding to the combination (T_1, P_1)? The experimenter was unable to study this particular combination. Moreover, the experimenter failed to study the interaction effect (if it exists) between these two factors. Interaction between two factors exists, if the effect of one factor on the output depends on the levels of the other factor. In the present example, if there is no interaction between temperature and pressure, then the output graphs at different levels of pressure (i.e. at P_0 and P_1) will be parallel. Non-parallel lines indicate interaction between the two factors. The interaction graph for the present example is shown in Figure 3.1. Here the lines are non-parallel and therefore we can conclude that there is an interaction between temperature and pressure, i.e. the effect of temperature on the yield of the process depends on the levels of pressure (or vice versa).

The following are the limitations of the one-factor-at-a-time approach to experimentation:

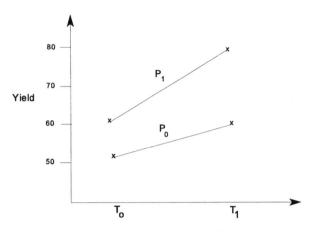

Figure 3.1. Example of an Interaction Graph

1. The one-factor-at-a-time method does not produce satisfactory results in a wide range of experimental settings.
2. Interaction among the factors under consideration for the experiment cannot be determined and therefore there is no guarantee that the final set of operating conditions will be even near the optimum.
3. This method is time consuming, inefficient and costly.

For complex manufacturing processes in today's industrial environment, interactions play an important role and therefore should be studied for achieving sound experimental conclusions. A better experimental strategy calls for the use of a **factorial experiment** [2, 3] in which all factors are varied simultaneously. Factorial experiments can be of two types: full factorial and fractional factorial experiments. A full factorial experiment allows experimenters to study all possible combinations of factors at their respective levels. For full factorial experiments, the experimenter may be able to vary all factors simultaneously and therefore permit the evaluation of interaction among the factors under study. A full factorial experiment can be denoted by (ℓ^k), where ℓ is the number of levels of the factor and k is the number of factors to be studied for the experiment. For example, 2^4 indicates that four factors are to be studied for an experiment in 16 experimental trials, provided all factors are at 2 levels. In many practical situations, however, these full factorial experiments require a prohibitively large number of runs, as the number of factors involved in the experiment increases. Therefore, the cost of performing full factorial experiments is exorbitantly high. It is good practice to perform a full factorial experiment, when the number of factors is less than or equal to 4. The question that now arises is "what happens if the number of factors is more than 4?" In such circumstances, **fractional factorial experiments** are recommended and can be used very effectively.

In fractional factorial experiments, we utilize only a fraction of the total size of the experiment. A fractional factorial experiment can be denoted by $\ell^{(k-p)}$, where $1/\ell^p$ is the fraction of the full factorial ℓ^k. For example, $2^{(6-2)}$ indicates that an experimenter wants to study six factors in just 16 experimental trials, provided all factors are at two levels. It would have required 64 runs to study six factors at two levels, if the experimenter had chosen a full factorial experiment. Because of the limited time and experimental resources, the experimenter has chosen one-fourth of the full factorial 2^6, by sacrificing information on interactions among some factors.

One of the major drawbacks of using such fractional factorial experiments is that experimenters may not be able to obtain a reliable and independent estimate for main and interaction effects. So one should be very cautious when selecting a fractional factorial experiment. Generally, sound engineering knowledge of the process and statistical skills are required to overcome this problem. The experimental designs employed by Taguchi, known as orthogonal arrays, are essentially fractional factorials [4]. We will explain the orthogonal array (OA) designs used by Taguchi for studying the effect of factors on the output performance of both products and manufacturing processes.

3.2   What are orthogonal arrays?

Orthogonal arrays are simple and useful tools for planning industrial experiments. An orthogonal array (OA) is a matrix of numbers arranged in rows and columns. Each row represents the levels (or states) of the selected factors in a given experiment, and each column represents a specific factor whose effects on the output (or response) are of interest to the experimenters. Using an orthogonal array, an experiment plan can be constructed easily by assigning factors to columns of the OA, then matching the different symbols of columns with the different factor levels. Orthogonal arrays have the balanced property that every factor setting occurs the same number of times for every setting of all other factors in the experiment. Orthogonality allows one to estimate the effects of each factor independently of the others. For example, a four-run OA can be used for studying two factors, say, A and B; each factor kept at two levels. Here there are four combinations of factor levels for A and B, i.e. A_1B_1, A_1B_2, A_2B_1 and A_2B_2. Table 3.4 illustrates a four-run OA for studying two factors at two levels.

Table 3.4.   A Four Run OA for Studying Two Factors

Trial number	Factor A	Factor B
1	1	1
2	1	2
3	2	1
4	2	2

Taguchi's OA tables are available for factors at two levels, three levels, mixed levels of two and three, and so on [5]. An orthogonal array is usually represented by the following notation:

$$L_a(b^c)$$

where a is the number of experimental trials required for the experiment; b is the number of levels of each factor in the experiment; and c is the number of columns in the array. The "L" notation implies that the information is based on the Latin square arrangement of factors. A Latin square arrangement is a square matrix arrangement of factors with separable factor effects [6]. In brief, the "L" notation indicates that the information is an orthogonal array. The number of columns indicates the number of factors that can be studied for a certain experiment.

For example, L_4 (2^2) can be used for studying two two-level factors in four experimental trials (or runs). As another example, L_9 (3^4) can be used for studying four three-level factors in nine runs. This book is focused on Taguchi's two-level orthogonal arrays, as research has shown that factors at two levels are the most widely used experiments in industries [7–9]. Moreover, factors at three and higher levels can easily discourage engineers and managers at the outset. This is because of the complexity inherent in the design and the mathematical skills required for the analysis and interpretation of such experiments. It is always a good practice to use factors at two levels as a teaching tool and to use the knowledge gained from this as a building block to study factors at higher levels. The standard OA tables for factors at two levels such as L_4, L_8, L_{12} and L_{16} are shown in Appendix A.

Dr Taguchi recommends that experiments should be designed with emphasis on main effects, to the extent possible. Past engineering experience should be used for the selection of output characteristics (or responses) with minimal interactions wherever possible. L_{12}, L_{18}, L_{36} and L_{54} arrays are among a group of specially designed arrays that enable the experimenters to focus on main effects. For these arrays, the interactions among the factors are distributed more or less uniformly to all columns in the array. L_4, L_8, L_{16}, etc. have been used widely by practitioners when interactions are to be studied for the experiment.

3.3 The role of orthogonal arrays

Matrix experiments using orthogonal arrays play a crucial role in seeing whether interactions are large compared to the main effects. Taguchi considers the ability to detect the presence of interactions to be the primary reason for using orthogonal arrays. To minimize financial resources, and due to time constraints, the experimenter usually attempts to employ the smallest size OA, which will meet the objective of the experiment. However, to test the validity of the additivity assumption, sometimes one uses a larger OA, which allows the evaluation of interaction among the factors under consideration, in addition to the main effects. Here additivity refers to an approximate representation of a cause–effect phenomenon in which one assumes the effects of factors on the output (or response) to be separable and independent of each other so that they are added together to compute the total effect of all factors present in the experiment [10]. One assumes that no interaction effects are present when one assumes additivity.

Orthogonal arrays (OAs) can lead to tangible savings while performing industrial experiments, but running experiments using OAs may not always suffice. Suppose an experimenter wishes to study four factors, say A, B, C and D, each factor kept at two levels (i.e. levels 1 and 2 respectively). A full factorial experiment would require 16 trials to study all the main and interaction effects. For industrial designed experiments, usually main effects and two-factor interactions (or two-way interactions) are to be studied for optimization problems. There are some situations in which three-factor interactions can be significant and therefore should be studied for satisfactory conclusions and further investigation [11, 12]. Table 3.5 displays a full factorial experiment structure for studying all the main factor effects, A, B, C and D, and all the possible interactions, orthogonal to each other. The observed output values are represented by $y_1, y_2, \ldots\ldots\ldots y_{16}$. In Table 3.5, y_1 is the response corresponding to the following factor-level combinations: factor A, level 1; factor B, level 1; factor C, level 1; and factor D, level 1. Similarly y_2 is the response corresponding to the following factor level combinations: factor A, level 1; factor B, level 2; factor C, level 1; and factor D, level 1, and so on. In order to support Table 3.5, a tree diagram (Figure 3.2) is constructed which enumerates all combinations for four factors at two levels.

Table 3.5. A 2^4 full factorial experiment with four factors A, B, C and D

		A_1		A_2	
		B_1	B_2	B_1	B_2
C_1	D_1	y_1	y_2	y_3	y_4
C_1	D_2	y_5	y_6	y_7	y_8
C_2	D_1	y_9	y_{10}	y_{11}	y_{12}
C_2	D_2	y_{13}	y_{14}	y_{15}	y_{16}

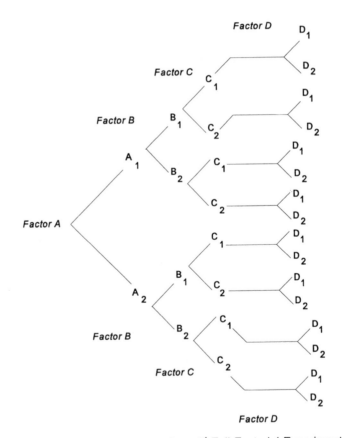

Figure 3.2.   Tree Diagram for a 2^4 Full Factorial Experiment

Now assume that the experimenter cannot afford to run the whole eight trials due to the limited time and resources allocated for the experiment. Under these circumstances, fractional factorials (or Taguchi's OAs) can be of great value to the experimenters. The experimenter has decided to perform a $2^{(4-1)}$ or Taguchi's L_8 OA to study all the main effects and two-factor interactions. Cells R-1 through R-8 indicated the eight experimental runs defined by Taguchi's L_8 OA for the experiment. Table 3.6 displays the experimental structure to study the four factors A, B, C and D, using an eight-run fractional factorial or OA experiment.

Table 3.7 shows an L_8 OA to support the above experimental layout. Here factor D is generated by assigning $D = ABC$, a third-order interaction (i.e. column 7 in L_8 OA). In Table 3.7, the numbers in parentheses represent the column numbers in the OA. The numbers in bold indicate the experimental run based on the combinations of factor levels of A, B, C and

Table 3.6. Experimental layout to study four factors using an L_8 OA

		A_1		A_2	
		B_1	B_2	B_1	B_2
C_1	D_1	R-1	X	X	R-7
C_1	D_2	X	R-3	R-5	X
C_2	D_1	X	R-4	R-6	X
C_2	D_2	R-2	X	X	R-8

Table 3.7. Taguchi's L_8 Orthogonal Array

Run	A (1)	B (2)	AB (3)	C (4)	AC (5)	BC (6)	D = ABC (7)
R-1	1	1	1	**1**	1	1	**1**
R-2	1	1	1	**2**	2	2	**2**
R-3	1	2	2	**1**	1	2	**2**
R-4	1	2	2	**2**	2	1	**1**
R-5	2	1	2	**1**	2	1	**2**
R-6	2	1	2	**2**	1	2	**1**
R-7	2	2	1	**1**	2	2	**1**
R-8	2	2	1	**2**	1	1	**2**

The two-factor interaction AB in column 3 is obtained by simply multiplying the factor levels of A and B in columns 1 and 2, respectively. The three-factor interaction ABC in column 7 is obtained by multiplying the factor levels of columns in 1, 2 and 4, respectively.

D. There are two discrepancies with this approach taken by the experimenter. First of all, the experimenter blindly believes that the three-factor or three-way interaction is negligible. Moreover, it would not be possible to study interactions between AB and CD, as interaction AB is confounded (or mixed) with interaction CD. This will not enable the experimenter to get an independent and reliable estimate of either interactions AB or CD. The term **confounding** refers to the combining influences of two or more factor (or interaction) effects in one measured effect. In other words, one cannot estimate factor effects and their interaction effects independently. This is one of the limitations of using fractional factorial or highly fractionated designs such as OAs. The confounding nature of factors among different columns in an OA can be obtained from Taguchi's interaction table (section 3.4). However, Taguchi's OAs are widely used in industries with the purpose of identifying factors that have large and significant effects on output product characteristics or process performance. A major use of these OAs is in screening the most important factors from a large number of factors in the early stages of any experimentation process [13]. The factors that are identified as significant are then investigated more thoroughly in subsequent experiments.

3.4 Linear graphs

A linear graph is a graphical tool developed by Taguchi to facilitate the assignment of factors and their interactions to an orthogonal array. The two important elements of a linear graph are "dots" (or nodes) and "lines". A "dot" on the linear graph represents a main factor effect and the "line" connecting two dots represents the interaction between the two corresponding factors [14]. In a linear graph, each dot and each line has a distinct column number(s) associated with it. When two dots are connected by a line, it means that the interaction of the two columns represented by the dots is confounded with the column represented by the line. The linear graphs for the OAs L_4, L_8 and L_{16} are illustrated in Appendix A.

The linear graph of the simplest OA (L_4) is shown in Figure 3.3. One may be able to study two main factor effects; say A and B (each factor at

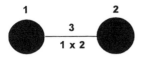

Figure 3.3.   Standard Linear Graph for L_4

two levels), and the interaction between them (i.e. AB or A × B). If the experimenter wants to study another factor effect (say, C at two levels) instead of the interaction AB, then again an L_4 would be suitable. In this particular example, factor A can be assigned to column 1, factor B to column 2 and either the interaction AB or factor C to column 3. If three factors are to be studied using an L_4 OA, then it is very important that factors A and B, which have already been assigned to columns 1 and 2 in the OA, do not interact with each other. This is because we assign C = AB as illustrated in Table 3.8. Hence the interaction of columns 1 and 2 is confounded with column 3.

It is very interesting to note from Table 3.8 that if the experimenter wants to study three factors, then the interaction between the first two factors (i.e. A and B) is confounded with the third factor C. The anomaly with this approach is that if column 3 is found to be significant from the statistical analysis, then the experimenter may fail to conclude whether column 3 is significant due to the main effect, C, or AB interaction. It is strongly advised to study three factors using an L_4 OA, if the experimenter is only interested in studying main effects. Also, it is important to ensure that the factors assigned in the columns of the array do not interact with each other. Generally, a sound engineering knowledge of the process, experience gained from previous experiments (if conducted) and some statistical skills are essential to make a valid decision.

Now assume that an experimenter wishes to study five factors, A, B, C, D and E, using an L_8 OA. The experimenter also wishes to study the interaction between B and C (i.e. BC). The L_8 OA is shown in Table 3.9.

For the present example, the experimenter may assign factor A to column 1, factor B to column 2, factor C to column 4, factor D to column 3 and factor E to column 5. The interaction between B and C (i.e. BC) is assigned to column 6. The standard linear graph for the L_8 OA is shown in Figure 3.4. The linear graph shows that the main effects to be studied should

Table 3.8.   An L_4 Orthogonal Array

Experimental trial (or run)	Factor A (1)	Factor B (2)	AB interaction or Factor C (3)
1	1	1	1
2	1	2	2
3	2	1	2
4	2	2	1

Table 3.9. L$_8$ Orthogonal Array to study five main effects and one interaction effect

Run/Trial	A (1)	B (2)	D = AB (3)	C (4)	E = AC (5)	BC (6)	ABC (7)
1	1	1	1	1	1	1	1
2	1	1	1	2	2	2	2
3	1	2	2	1	1	2	2
4	1	2	2	2	2	1	1
5	2	1	2	1	2	1	2
6	2	1	2	2	1	2	1
7	2	2	1	1	2	2	1
8	2	2	1	2	1	1	2

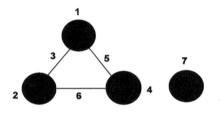

Figure 3.4.   Standard Linear Graph for L$_8$

be assigned to columns 1, 2, 4 and 7 of the L$_8$ OA. The interaction between columns 1 and 2 should be assigned to column 3, that between columns 2 and 4 to column 6; and that between columns 1 and 4 to column 5. Therefore columns 3, 5 and 6 represent the interaction effects to be studied using this array.

The linear graphs advocated by Taguchi suffer from the problem that they do not provide the complete confounding relationships (especially for highly fractionated designs) among the factors or interactions to be studied for the experiment. In other words, a linear graph does not show the interaction between every pair of columns of the OA [15]. Therefore one may not be able to determine the true effect of a factor or its interaction with other factors. There are two alternatives to rectification of this problem. The first alternative is to use the **triangular table of interactions** or **interaction table**. A triangular table provides information about the interaction of the various columns of an OA. The triangular tables of interactions for L$_4$, L$_8$ and L$_{16}$ are shown in Appendix A. The triangular table for the L$_8$ OA is shown in Table 3.10.

The interaction tables are generated directly from the linear algebraic

Table 3.10.   Interaction Table for L_8 OA

| | | | | Column | | | |
Column	1	2	3	4	5	6	7
1	(1)	3	2	5	4	7	6
2		(2)	1	6	7	4	5
3			(3)	7	6	5	4
4				(4)	1	2	3
5					(5)	3	2
6						(6)	1
7							(7)

relations that were used in creating the orthogonal arrays themselves. The standard linear graph of L_8 reveals that column 3 is the interaction between the factors assigned to columns 1 and 2. In fact, column 3 is the interaction between columns 1 and 2 in the L_8 OA, but also between columns 5 and 6 and columns 4 and 7. This deficiency is rectified in Table 3.10. The entries in the table show the column with which the interaction between every pair of columns is confounded. The table should be interpreted in the following manner. Consider column 4 in Table 3.10. The first element in column 4 (i.e. 5) is the interaction between columns 1 and 4 in the L_8 OA. The second element in column 4 (i.e. 6) is the interaction between 2 and 4 and so on. Now consider column 7 in the L_8 OA. From the interaction table for L_8, we can say column 7 is the interaction between columns 2 and 5, columns 1 and 6 and columns 3 and 4. This information cannot be obtained from the linear graph for L_8 OA. Readers with limited mathematical and statistical skills are recommended to follow the first alternative. Readers with reasonable mathematical and statistical skills are encouraged to follow both alternatives for better understanding of the aliasing (or confounding) structures of highly fractionated designs.

The second alternative is to consider the **resolution of the design**. Design resolution (R) is a summary characteristic of confounding structures or patterns. If a design is of resolution R, then any effect involving $s < R$ factors is not aliased with any other effect that involves fewer than $t = R - s$ factors [16]. For example, if a design has a resolution IV, then any main effect $(s = 1)$ is not confounded with any other effect involving fewer than $4 - 1 = 3$ factors.

Design resolution identifies for a specific design the order of confounding of the main effects and their interactions. It is a key tool for determining what OA or fractional factorial design will be the best choice for a problem under investigation. Generally speaking, design resolution measures the degree to which the main factor effects are confounded with the interaction effects (i.e. two-factor or higher interactions). In particular, designs of resolutions III, IV and V are most important, and are defined as follows:

Resolution III designs. These are designs in which no main effects are confounded with any other main effect, but main effects are confounded with two-factor (or two-way) interactions. Resolution III designs are widely used for screening experiments, where we separate out critical factors from a large number of factors as part of the initial investigation of a process. For example, consider an L_4 OA to study three factors, A, B and C (refer to Table 3.8). Here factor C assigned to column 3 is the interaction between factor A in column 1 and factor B in column 2. In other words, the interaction AB is confounded with the main effect C. Therefore we can write, C = AB, which is called the **design generator** of the design. Having obtained the design generator, the next step is to state the **defining relation** (or defining contrast) of the design. The defining relation generally produces the confounding pattern or aliasing structures. Effects that are confounded are called aliases. A list of the confoundings that occur in an experimental design is called an alias structure or a confounding pattern. The design generators for two-level OA or fractional factorial designs for the number of factors varying from three to seven are shown in Appendix B.

The defining relation for the above design is obtained by multiplying both sides of C = AB by C. Mathematically, this can be written as $C \times C = AB \times C$. This reduces to I = ABC, where "I" is called the identity element. It is very interesting to note that the number of factors in the defining relation will yield the resolution of the design. Here there are three factors (i.e. A, B and C) in the defining relation and therefore the resolution of the design is III. The aliasing structure can now simply be obtained as follows:

I = ABC (defining relation of the design)

$I \times A = A \times ABC \Rightarrow A = BC$ ($A \times A = A^2 = 1$, $I \times A = A$).

Similarly, $I \times B = B \times ABC \Rightarrow B = AC$ ($B \times B = B^2 = 1$, $I \times B = B$);

$I \times AB = AB \times ABC = C$, and so on.

The full aliasing structure or confounding pattern of the design is shown in Table 3.11.

Table 3.11 tells us that column 1 in an L_4 OA is confounded with columns 2 and 3, column 2 is confounded with columns 1 and 3, and column 3 is

Table 3.11. Confounding Pattern of the L_4 OA design

I = ABC *(Defining relation)*

A = BC	B = AC	C = AB
AB = C	AC = B	BC = A

Table 3.12. Confounding Pattern of the L_8 OA to study five factors

I = ABD = ACE = BCDE *(Defining relation of the design)*

Confoundings of main effects
A = BD = CE = ABCDE   B = AD = ABCE = CDE   C = ABCD = AE = BDE

Confoundings of two-factor interaction effects
AB = D = BCE = ACDE   AC = E = BCD = ABDE   BC = ACD = ABE = DE

Confounding of three-factor interaction effect
ABC = CD = BE = ADE

confounded with columns 1 and 2. For these designs, the main effects can be estimated independently only if the two-factor interactions are not significant.

Now consider another situation where the experimenter wishes to study five factors A, B, C, D and E using the L_8 OA. The design generators for this design are given by: D = AB and E = AC. Here main effect, D, is confounded with interaction AB and main effect, E, is confounded with interaction AC. The defining relations are therefore given by: I = ABD and I = ACE. Combining the defining relations will yield:

$$I = ABD = ACE = BCDE \text{ (Note: BCDE = ABD} \times \text{ACE).}$$

In such situations, the resolution of the design is given by the smallest number of factors in any of the defining relations, including the implicit ones. In the present example, the resolution of the design is III, because the smallest number of factors in the defining relations is three. The complete confounding pattern of the design is shown in Table 3.12. For industrial experiments, we need to scrutinize which of the main factors and interaction effects are confounded with other interaction effects. Factor interactions up to second order should be examined for obtaining satisfactory conclusions. All higher-order interactions can be neglected as they have very little influence compared to main and two-factor interactions. However, there are few situations where three-factor interactions are statistically significant [11, 12].

Resolution IV designs. These are designs in which main effects are not confounded with other main effects or two-factor interactions; but main effects are confounded with three-factor interactions, or two-factor interactions are confounded with other two-factor interactions. Therefore the main effects can be estimated independently regardless of the significance of the two-factor interactions. For example, an experimenter wishes to study four factors A, B, C and D using an L_8 OA. This is very similar to a $2^{(4-1)}$ factorial experiment. The design generator for this design is given by: D = ABC (or 7 = 124, where 1, 2, 4 and 7 are columns in the OA). Here the main factor effect, D, is confounded with a three-factor interaction (ABC). The defining relation of the design is given by: I = ABCD. The number of factors in the defining relation is four, and therefore the resolution of the design is IV. The full confounding pattern of the design is given in Table 3.13.

Table 3.13 tells us that columns 1, 2, 4 and 7 are confounded with three-factor interactions and columns 3, 5 and 6 are confounded with two-factor interactions.

Resolution V designs. These are designs in which main effects are not confounded with other main effects, two-factor interactions or three-factor interactions; but main effects are confounded with four-factor interactions, or two-factor interactions are confounded with three-factor interactions and vice versa. These designs are almost as powerful as that of full factorial designs. For example, study of five factors A, B, C, D and E using an L_{16} OA or $2^{(5-1)}$ is a resolution V design. The design generator for the design is given by: E = ABCD. The defining contrast of this design is: I = ABCDE. The number of factors in the defining contrast is five and therefore the resolution of the design is five. The full confounding pattern of the design is shown in Table 3.14.

Table 3.14 reveals that all main effects are confounded with four-factor interactions and all two-factor interactions are confounded with three-factor interactions.

Table 3.13. Confounding Pattern of the L_8 design to study four factors

---

I = ABCD *(Defining relation or contrast)*

Confoundings of main effects
A = BCD   B = ACD   C = ABD   D = ABC

Confoundings of two–factor interaction effects
AB = CD   AC = BD   AD = BC

Table 3.14. Confounding Pattern of the L_{16} OA design to study five factors

$I = ABCDE$ *(Defining contrast of the design)*				
Confoundings of main effects				
$A = BCDE$	$B = ACDE$	$C = ABDE$	$D = ABCE$	$E = ABCD$
Confoundings of two-factor interaction effects				
$AB = CDE$	$AC = BDE$	$AD = BCE$	$AE = BCD$	$BC = ADE$
$BD = ACE$	$BE = ACD$	$CD = ABE$	$CE = ABD$	$DE = ABC$

From the above discussion of design resolution, the following comments can be made:

1. Higher-resolution designs seem to be more desirable as they provide an opportunity for main effect and two-factor interaction effect estimates to be determined in an unconfounded state, assuming that higher-order interaction effects (generally three or more) can be neglected.

2. There is a limit to the number of variables or factors that can be studied in a fixed number of experiments while maintaining a pre-specified resolution requirement. For example, if an experimenter wishes to perform an eight-run experiment using L_8 OA and would like to maintain resolution IV for the experiment, then the maximum number of variables that can be tested by the experimenter must be four.

3. Resolution III designs are commonly referred to as saturated designs. A design is said to be saturated if $(c - 1)$ factors are to be studied in a runs. Note that an OA is represented by L_a (b^c). For example, studying three factors at two levels using an L_4 OA, seven factors at two levels using an L_8 OA, and so on.

3.5 Degrees of freedom

In statistics, "degrees of freedom" is the number of observations that can be varied independently of each other. It is often represented by v. The number of degrees of freedom depicts the number of fair and independent comparisons that can be made in a set of data. For example, consider the weight of three students, say X, Y and Z. If the weight of X is W_x and that of Y is W_y, then we can make one fair comparison $(W_x - W_y)$. If the weight of Z is W_z, then we can make another fair comparison $(W_x - W_z)$. The comparison between W_y and W_z is not fair because:

$$W_y - W_z = (W_x - W_z) - (W_x - W_y).$$

In the context of experimental design, the number of degrees of freedom associated with a factor or process variable is equal to one less than the number of levels for that factor. Throughout this book, the number of degrees of freedom associated with a factor is represented by v_f. For example, suppose a chemist wants to study the effect of a certain catalyst on the yield of the chemical process. He decides to choose three types of catalyst, say catalyst A, catalyst B and catalyst C. Here the factor "catalyst" has three levels and therefore the number of degrees of freedom associated with it is equal to 2 (i.e. number of levels − 1). Now consider another example: an engineer wishes to study the effect of barrel temperature, studied at two levels, say $240°C$ and $280°C$ on a certain injection-moulded part. Here the factor "temperature" was kept at two levels and therefore the number of degrees of freedom associated with it is equal to 1.

The number of degrees of freedom of an OA (v_{OA}) is equal to one less than the number of experimental trials, because one degree of freedom is always taken up or possessed by the overall mean. For example, an L_8 OA has seven degrees of freedom as the number of experimental trials associated with the array is equal to 8. Therefore we may be able to study up to seven factors, each factor kept at two levels. Therefore the choice of an OA design depends on the degrees of freedom associated with each of the factors to be studied for the experiment.

In order to use a standard OA, the number of degrees of freedom associated with factors at their respective levels should be matched with the number of degrees of freedom for that OA. For example, consider the L_{12} OA to study 11 factors at two levels (i.e. L_8 (2^{11})). The number of degrees of freedom for the OA (v_{OA}) is equal to 11. This is the case when there are no repetitions or replications of experimental trials. The total number of degrees of freedom (v_{ft}) associated with the factors is given by:

$$V_{ft} = (number\ of\ levels - 1) \times (number\ of\ factors\ at\ that\ specified\ level)$$

$$= (2 - 1) \times 11 = 11. \tag{3.1}$$

This implies that we cannot study more than 11 two-level factors using the L_{12} OA. It is very important to note that, v_{OA} should be always greater than or equal to v_{ft}.

Repetition. In repetition, an experimenter may repeat an experimental trial (or run) as planned, before proceeding to the next trial in the experimental layout. The experimental trial sequence is selected in a random order. For example, given the trial sequence 3, 1, 4 and 2 of an L_4 OA, two successive runs of trial number 3 are made followed by two successive runs of trial number 1 and son on. The advantage of this approach is that the experimental set-up cost should be minimum. However, a set-up error is unlikely to be identified or detected. Moreover, the effect of external factors, such as humidity, dust, vibration, etc., may not be captured during the successive runs, if the time to complete each run is short.

Replication. Replication is a process of running all the experimental trials in a random order. One way to decide the order is to pull one trial number at a time randomly from a set of trial numbers, or to use a random generator. Results from replication contain more information than those from repetition as replication captures variation in results due to experimental set-up. Since replication requires resetting of each trial condition, the cost of the experiment will be increased to some extent. The idea of using this concept is to reduce the effect of undesirable factors induced in experiments.

Suppose an experimenter wishes to study three factors using an L_4 OA. Now assume that each of the four trials is replicated twice. In this case, the total degrees of freedom for the experiment is given by:

$$\nu_{\text{experiment}} = (rn) - 1 \tag{3.2}$$

where r = number of repetitions or replications, and n = number of trials in the array. Using equation 3.2, we obtain $\nu_{\text{experiment}} = (2 \times 4) - 1 = 7$.

This does not mean that one may be able to study seven factors at two levels. But the total degrees of freedom in this case is the sum of the degrees of freedom for main effects and error. This error degrees of freedom will be used for performing statistical analysis and will be discussed in Chapter 9.

So far we have been discussing experiments where the interactions among the factors were not to be investigated for the experiment. The calculation of the total degrees of freedom when interaction effects are to be studied is illustrated as below.

The degrees of freedom for an interaction is the product of the degrees of freedom of each individual factor involved in the interaction. Consider two factors A and B. Let ν_A and ν_B be the number of degrees of freedom for A and B respectively. The degrees of freedom for interaction between A and B (i.e. AB) is given by: $\nu_{AB} = \nu_A \times \nu_B$, where ν_{AB} is the number of degrees of freedom for interaction AB. For example, assume that factors A and B are at two levels and the experimenter wishes to study both main factor effects due to A and B and also interaction AB, using an L_4 OA.

Here the total degrees of freedom for the experiment is 3. The number of degrees of freedom for factors A and B is 1 (i.e. $v_A = 1$ and $v_B = 1$). Therefore the number of degrees of freedom for interaction AB (v_{AB}) is equal to 1.

Now consider another example where an experimenter wishes to study four three-level factors (A, B, C and D) and one two-level factor (H). The experimenter would also like to study the interaction between A and C as well as B and H. How many degrees of freedom are required for this experiment? The number of degrees of freedom for a three-level factor is 2. As there are four factors at three levels, the number of degrees of freedom required is $4 \times 2 = 8$. The number of degrees of freedom for a two-level factor is equal to 1. The number of degrees of freedom required for studying interaction AC (v_{AC}) is equal to $v_A \times v_C = 2 \times 2 = 4$. Similarly, the number of degrees of freedom required for studying interaction BH (v_{BH}) is equal to $v_B \times v_H = 2 \times 1 = 2$. Therefore the total degrees of freedom required for the experiment is equal to $8 + 1 + 4 + 2 = 15$.

3.6 Randomization in industrial designed experiments

While designing industrial experiments, there are always factors, such as power surges, operator errors, etc., which may influence the outcome but are not included in the experimental design because they are too difficult to control, or because they are unknown. Such factors can substantially bias the experimental results. One simple way for reducing the effects of bias is the use of randomization. Using the concept of randomization, one can distribute the bias effects evenly over all treatment combinations in the experiment. In other words, randomization can ensure that all levels of a factor have an equal chance of being affected by external sources of variation. Nelson advocates "look upon randomisation as insurance, and buy as much of it as you can afford" [17]. Shainin accentuates the importance of randomization for industrial designed experiments as "experimenter"s insurance policy". According to Shainin "failure to randomise mitigates the statistical validity of an experiment" [18]. On the other hand, it might be completely impractical to randomize, as in the case of a large heated reactor being studied at three temperatures. By going from the lowest through the middle to the highest temperature, one may complete the experiment in one day. But suppose randomization required the sequence to be high, low and then middle heat; to cool down from the high heat to the low heat might take up to a week! Because of the demands of time, cost and effort required to change the factor levels of some factors in an experiment, it might be desirable to change the factor levels less frequently than others. In such situations, **restricted randomization** can be accommodated.

Restricted randomization provides adequate protection against unknown sources of bias without imposing burdensome constraints on the conduct of the experiment. Some experimenters believe in running the experimental trials in a systematic or standard order, and using engineering knowledge and skills to identify the unknown sources of variation [19].

3.7 Selecting a standard OA for two-level factors

In order to select a standard OA for an experiment, the following inequality must be satisfied.

$$v_{OA} \geq v_{\text{required for main effects and interactions}}.$$

Prior to selecting a standard OA, we have to calculate the total degrees of freedom required for the experiment and then compare this with the degrees of freedom for the nearest OA. If only main effects are of concern to the experimenters, then Table 3.15 can be of great value to select the most appropriate OA for a given problem.

Example 1. An experimenter has identified three two-level process variables (or factors) for a certain injection-moulding process. The experimenter was only interested in studying the main factor effects. What is the suitable OA for this study?

Since a two-level factor has one degree of freedom, three two-level process variables would require three degrees of freedom. Therefore the smallest

Table 3.15. Selecting a Standard Orthogonal Array for factors at 2-levels

Type of OA	Number of Experimental Trials	Maximum number of main effects that can be studied
L_4	4	3
L_8	8	7
L_{12}	12	11
L_{16}	16	15
L_{32}	32	31

OA that can be used should have at least three degrees of freedom. Table 3.15 shows that the smallest OA with three degrees of freedom is the L_4 OA. Therefore the most suitable OA for this study is an L_4 OA.

Example 2. An engineer wishes to study two two-level factors and the interaction between them for a certain casting process. What is the acceptable OA for this study?

Here the two two-level factors would require two degrees of freedom. The number of degrees of freedom required for the interaction between them is equal to 1 (i.e. $1 \times 1 = 1$). Therefore the total degrees of freedom required for studying two factor effects and the interaction between them is equal to 3. The most acceptable OA for this experiment would be an L_4 OA.

Example 3. An injection-moulding process engineer has been experiencing problems associated with increased parts shrinkage variability which occurs after curing. Seven factors at two levels were identified for the study. The engineer was concerned only about main factor effects as part of initial investigation of the study. What should be the most appropriate OA for this study?

Since a two-level factor has one degree of freedom, seven factors at two levels would need seven degrees of freedom. In this case, the smallest OA that can be used should have at least seven degrees of freedom. Table 3.15 shows that the most suitable OA for achieving the above objective is the L_8 OA.

Example 4. In a cake-baking experiment designed to determine the best recipe for a pound cake, five factors at two levels were identified. Among these factors, two two-factor interactions were of interest to the experimenter for study. What should be the acceptable OA for this study?

This example can be best illustrated as follows:

Degrees of freedom for studying five main effects = 5.

Degrees of freedom for studying two two-factor interactions = 2.

Total degrees of freedom = 7.

From Table 3.15, the most acceptable array for this study is an L_8 OA.

Example 5. In a certain arc-welding process, an experimenter wishes to study six factor effects and two interaction effects for maximizing the weld strength. The experimenter has kept each factor at two levels. What is the best choice of OA for this particular study?

Degrees of freedom required for the main effects = 6.

Degrees of freedom for studying two interaction effects = 2.

Total degrees of freedom = 8.

The smallest OA that can be used for this study should have at least eight degrees of freedom. Table 3.15 shows that the smallest OA that can match this objective is the L_{12} OA. But an L_{12} OA is not suitable when interactions among factors are to be studied. The next smallest OA is the L_{16} OA. Therefore the best choice of array for this study is an L_{16} OA.

Example 6. An engineer wants to improve the breakage strength and porosity of airframe parts from a pressure diecasting process. Seven factors and five interactions were chosen for study, based on results from previous experiments. Each factor is to be evaluated at two levels. Factor levels were chosen from previous experience. What should be the most appropriate array for this investigation?

Degrees of freedom for seven main factor effects = 7.

Degrees of freedom required for studying five interaction effects = 5.

Total degrees of freedom required for the experiment = 12.

The best choice of array for this investigation is an L_{16} OA.

Example 7. An experimenter wishes to improve solderability and thereby reduce the solder defects on printed circuit boards (PCBs). In order to achieve the above objective, it is important to understand the factors that influence solderability. Eleven factors were identified to be important for the study. Each factor is to be evaluated at two levels. As part of initial investigation, the experimenter wants to study only the main factor effects. What should be the most suitable array for this investigation?

As the experimenter is interested only in main effects, the number of degrees of freedom for studying 11 main effects is 11. From Table 3.15, the most suitable array for accomplishing this objective is an L_{12} OA.

Example 8.   In a certain company, an engineer wants to identify the causes of automotive shock-absorber damping force variability. Seven factors and five interactions are of interest to the engineer. Each factor is to be studied at two levels. What should be the most appropriate choice of OA for this study?

Degrees of freedom for main effects = 7.

Degrees of freedom for interaction effects = 5.

Total degrees of freedom = 12.

The smallest OA which can be used for this study should have at least 12 degrees of freedom. Table 3.15 shows that the smallest OA with 12 degrees of freedom is the L_{16} OA. Therefore the most suitable OA for this study is an L_{16} OA.

Exercises

3.1   A plant manager has decided to design an experiment to determine the effects of two factors, temperature and pressure, on defect rate (i.e. response) of a certain product. Two levels for each factor were chosen for the experiment. The experiment was performed in two stages. In the first stage, he tested the product at temperatures T_1 and T_2, keeping the pressure constant at level P_0. The following results were obtained:

Pressure	Temperature	Response
P_0	T_1	0.032
P_0	T_2	0.016

Because a more desirable response was obtained at T_2, the plant manager set the temperature at T_2, and tested the product by changing the level of pressure from P_0 to P_1. The results are shown below.

Pressure	Temperature	Response
P_0	T_2	0.016
P_1	T_2	0.010

From the above results, the manager has arrived at a conclusion that the ideal setting to minimize the defect rate is: temperature T_2 and pressure P_1.

Discuss the problems of this experimentation approach taken by the plant manager.

3.2 Explain the role of orthogonal arrays in Taguchi experiments.

3.3 Suppose it is desired to study six factors at two levels in a full factorial design scheme.
   (a)   How many experimental trials are required?
   (b)   How many main factor effects can be estimated?
   (c)   How many two-factor interactions can be estimated?

3.4 An experimenter wishes to study the effect of five factors on the burr height in a metal-stamping process. Funds are available to conduct only eight trials; therefore it was decided to use an L_8 OA for this experiment.
   (a)   What generators should be used to get the highest resolution?
   (b)   What is the defining relation of the design?
   (c)   Show the alias structure of the design (assume that third- and higher-order interactions are negligible).
   (d)   Compare this with the interaction table.

3.5 Explain the term "degrees of freedom". Illustrate by an example that for four measurements, X_1, X_2, X_3 and X_4, one can make only three fair comparisons.

3.6 An experimenter wishes to study seven two-level factors and two second-order interactions. Calculate the degrees of freedom required for the experiment. Also suggest a suitable OA for the experiment. Show the alias structure and discuss the limitations of the design.

References

1. Juran, J.M. (1974) "Quality Control Handbook", McGraw-Hill Publishers, 3rd Edition.
2. Montgomery, D.C. (1991) "Design and Analysis of Experiments", John Wiley and Sons, Chichester.
3. Box, G.E.P. et al. (1978) "Statistics for Experimenters", John Wiley Publishers.
4. Lin, Dennis K. (1994) "Making Full Use of Taguchi's Orthogonal Arrays", International Journal of Quality and Reliability Engineering, Vol. 10, pp. 117–121.
5. Taguchi, G. and Konishi, S. (1987) "Orthogonal Arrays and Linear Graphs", ASI, Center for Taguchi Methods, ASI Press.
6. Belavendram, N. (1995) "Quality by Design: Taguchi Techniques for Industrial Experimentation", Prentice-Hall Publishers.
7. Gunst, R.F. et al. (1991) "How to Construct Fractional Factorial Experiments", ASQC Statistics Division, ASQC Press.

8. Ross, J. (1988) "Taguchi Techniques for Quality Engineering", Mcgraw-Hill Publishers.
9. Lucas, J.M. (1992) "Split Plotting and Randomisation in Industrial Experiments", ASQC Quality Congress Transcations, Nashville, pp. 374–382.
10. Bagchi, T.P. (1993) "Taguchi Methods Explained—Practical Steps to Robust Design", Prentice-Hall of India.
11. Heil, H. and Horma, B. (1986) "Maximisation of Weld Strength Using Taguchi Methods, 4th Symposium on Taguchi Methods, pp. 273–287.
12. Antony, J. and Kaye, M. (1996) "Optimisation of Core Tube Life Using Taguchi Experimental Design Methodology", Quality World Technical Supplement, March pp. 42–50.
13. Montgomery, D.C. (1991) "Design and Analysis of Experiments", John Wiley and Sons, Chichester.
14. Anand, K.C. et al. (1994) "Wet method of Chemical Analysis of Cast Iron: Upgrading accuracy and precision through experimental design", Quality Engineering, ASQC, pp. 187–208.
15. Tsui, K-L. (1988) "Strategies for planning experiments using Orthogonal Arrays and Confounding Tables", Quality and Reliability Engineering International, Vol. 4, 113–122.
16. Mason, R.L., Gunst, R.F., and Hess, J.L. (1989) "Statistical Design and Analysis of Experiments", John Wiley and Sons, New York.
17. Nelson, L.S. (1996) "Notes on the Use of Randomisation in Experimentation", Journal of Quality Technology, Vol. 28, No. 1, pp. 123–126.
18. Shainin, R.D. (1993) "Strategies for Technical Problem Solving", Quality Engineering, Vol. 3, pp. 433–448.
19. Barker, T.B. (1990) "Engineering Quality by Design—Interpreting the Taguchi Approach", Marcel Dekker, Inc.

4 ASSIGNMENT OF FACTOR AND INTERACTION EFFECTS TO AN OA

4.1 Introduction

Orthogonal arrays have several columns available for assignment of main factors and some columns subsequently can estimate the effect of interactions of those factors. Note that all possible interactions between factors cannot be estimated from using the OA; a full factorial type of experimentation would be desirable to meet this objective. Identification of interactions prior to conducting the experiment need knowledge obtained from previous experiments or experience gained of the relevant process under investigation. If there are unnecessary columns in the OA, then after assigning the main factors whose main effects are desirable, those columns may be assigned to the possible interactions or interactions which are of interest to the experimenter for study. Linear graphs are useful to illustrate the assignment of main factors and the anticipated interactions among the factors. Because linear graphs do not show the interaction between every pair of columns of an OA, it is recommended to use the interaction table which provides the required information.

4.2 How to assign factor effects to an OA

The following guidelines may be useful while assigning main factors and interactions (if any) to an OA.

1. Calculate the total degrees of freedom required for main effects and interaction effects (if any) to be studied for the experiment.
2. Choose the most suitable OA which should match the degrees of freedom for the experiment. It is important to bear in mind that the degrees of freedom required for the OA should be greater than or equal to the total degrees of freedom required for studying the main and interaction effects.
3. Draw the problem graph. A problem graph is a graphical representation of a given problem [1, 2]. A dot in a problem graph represents a factor and a line denotes an interaction.
4. Select an appropriate standard linear graph. There may be many choices. Decide on the most suitable one.
5. Fit the problem graph to one of the standard linear graphs of the OA selected.
6. Assign each factor and interaction to the appropriate column.

Example 1.   In a certain milling operation, three factors are identified from brainstorming which might have an impact on the surface finish of the finished parts. The experimenter wants to study all factors at two levels using a Taguchi-style experiment. The list of factors and the interactions of interest are shown below.

Main effects: cutting speed (factor A); depth of cut (factor B); feed (factor C).

Interactions of interest: AC and BC.

What is the most suitable OA for this study and how does the experimenter assign factors and interactions into this array?

As the experimenter wishes to study three main and two interaction effects (each factor kept at two levels), the total degrees of freedom required for the experiment is equal to 5. Table 3.15 shows that the smallest OA which can match this objective is the L_8 OA. Having selected the OA, the experimenter has constructed a problem graph as shown in Figure 4.1.

Having constructed the problem graph, the next step is to compare this problem graph with the standard linear graph shown in Figure 3.4 (see Chapter 3). The objective here is to fit the problem graph into the standard

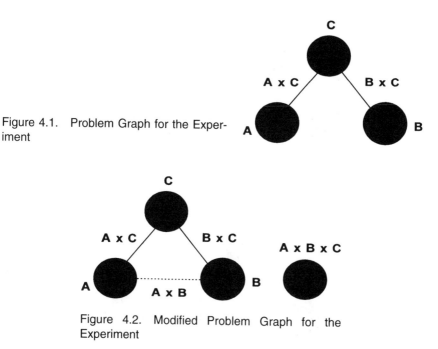

Figure 4.1. Problem Graph for the Experiment

Figure 4.2. Modified Problem Graph for the Experiment

linear graph of the OA selected. The modified problem graph is shown in Figure 4.2.

The interactions that are of interest to the experimenter are represented by solid lines and that which is not of interest is represented by a dotted line. The experimenter assigns main factor effects and interactions of interest into the L_8 OA in the following manner: column 1, factor C; column 2, factor A; column 3, interaction AC; column 4, factor B; column 5, interaction BC; column 6, interaction AB; and column 7, interaction ABC.

Example 2. An engineer is interested in improving the life of a cutting tool. Five factors, say A, B, C, D and E (each factor at two levels) are identified as affecting the life of the tool. Two interactions, BC and CD, are also of interest to the engineer for study. The engineer decides to use an L_8 OA for this study. How does he assign these main factor and interaction effects into the array?

The problem graph for this experiment is shown in Figure 4.3. Comparing Figure 4.3 with the standard linear graph for L_8 in Figure 3.4, the problem graph is modified to Figure 4.4. In this example, the experimenter was not interested in studying interaction BD. However, main effect A is

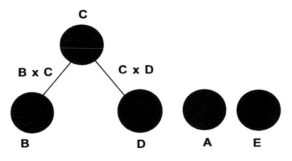

Figure 4.3.   Problem Graph for the Experiment

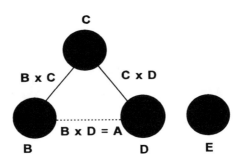

Figure 4.4.   Modified Problem Graph for
the Experiment

assigned to interaction column BD (see Figure 4.4). If a main effect is
assigned to an interaction column (though the experimenter has no interest to study the interaction), the situation is represented by a dotted line
and not by a solid line.

The assignment of main and interaction effects made by the experimenter
to various columns of the array are: column 1, factor C; column 2, factor B;
column 3, interaction BC; column 4, factor D; column 5, interaction CD;
column 6, factor A (A = BD); column 7, factor E (E = BCD = AC).

Example 3.   An experimenter is interested to study five factors (each factor
at two levels) and two interactions. The list of factors and interactions of
interest are as follows.

Main effects: A, B, C, D and E.

Two-factor interactions: AB and CD.

The experimenter has decided to use an L_8 OA. How does the experimenter study factors and interactions using the selected array? Discuss the problems of using an L_8 array for this study.

An L_8 OA has seven degrees of freedom. Therefore the experimenter may be able to study seven effects (if each factor is kept at two levels). If A and B are assigned to columns 1 and 2, then column 3 will be reserved for interaction AB. This means that columns 4, 5, 6 and 7 remain for factors C, D, E and interaction CD. However, any combination of these columns 4, 5, 6 and 7 contains only the values 1, 2 or 3, hence their interaction would involve columns previously assigned to A, B and interaction AB. The problem graph for this experiment is shown in Figure 4.5.

Interactions represented by solid lines are of interest and those represented by dotted lines are of no interest to the experimenter. Here the main effect, D, is confounded with interaction AB and E with interaction AC. This can be envisaged easily by obtaining the confounding structure of the design. The design generators are: D = AB and E = AC. The defining relation is given by: I = ABD = ACE = BCDE. The complete confounding structure for main and two-factor interaction effects is shown in Table 4.1.

The problem with this design is that the experimenter will not be able to obtain an independent estimate for interaction AB, as main effect D is confounded with it. Also, interaction CD is confounded with interaction BE. Therefore it is advisable not to choose an L_8 OA for this study. Moreover, an L_{12} OA is also not appropriate for this study as it is not suitable for studying the desired interaction effects AB and CD. Therefore it can be concluded that the most suitable array for this particular experiment is an L_{16} OA.

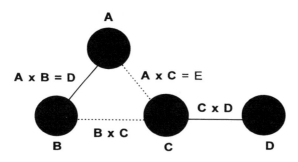

Figure 4.5. Problem Graph for the Experiment

Table 4.1.   Confounding Structure of the Design

Design Resolution of the design: III
Main Effects:
A = BD = CE = ABCDE   B = AD = ABCE = CDE
C = ABCD = AE = BDE   D = AB = ACDE = BCE
E = ABDE = AC = BCD

Two-factor interactions:
AB = D = BCE = ACDE   AC = BCD = E = ABDE
BC = ACD = ABE = DE   AD = B = CDE = ABCE
AE = BDE = C = ABCD   BD = A = ABCDE = CE
BE = ADE = ABC = CD   CD = ABC = ADE = BE
CE = ABCDE = A = BD   DE = ABE = ACD = BC

Example 4. How does the above experimenter assign the factor and interaction effects into an L_{16} OA?

There are so many ways to achieve this column assignment. The assignment of main and interaction effects can be made in the following manner:

column 1, factor A;	column 2, factor B;	column 3, interaction AB;
column 4, factor C;	column 5, interaction AC;	column 6, interaction BC;
column 7, interaction DE;	column 8, factor D;	column 9, interaction AD;
column 10, interaction BD;	column 11, interaction CE;	column 12, interaction CD;
column 13, interaction BE;	column 14, interaction AE;	column 15, factor E.

A modified problem graph for the experiment is shown in Figure 4.6.

Example 5. In a certain welding process, an experimenter wishes to study the effect of seven factors on the weld strength. The factors and interaction effects of interest are: A, B, C, D, E, F and G and five interaction effects: AB, AD, BD, BC and CE. Each factor needs to be evaluated at two levels for the initial study. It is decided to select an L_{16} OA for this study. How does the experimenter assign these factors and interactions to the selected OA?

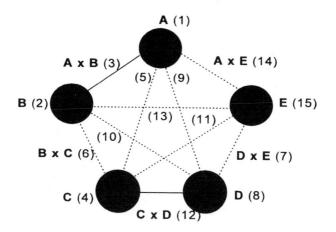

Figure 4.6.   Modified Problem Graph for the Experiment

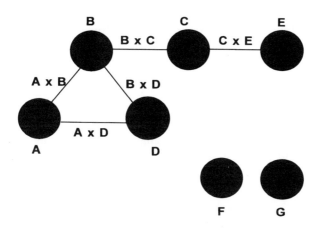

Figure 4.7.   Problem Graph for the Experiment

The first step is to construct a problem graph that illustrates the pictorial representation of the problem. The problem graph for the experiment is shown in Figure 4.7.

Having constructed the problem graph, the next step is to modify the problem graph to fit one of the standard linear graphs for the selected OA. The modified problem graph is shown in Figure 4.8. It is important to note that those effects that are of no interest to the experimenter in various columns of the OA may be treated as error, *e*. This error column in the

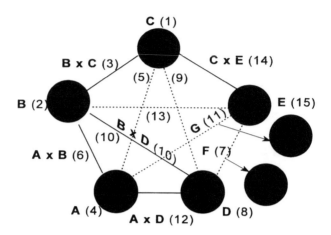

Figure 4.8.   Modified Problem Graph for the Experiment

OA can be of great value when there are insufficient degrees of freedom for error. The role of error and its interpretation will be explained in Chapter 9.

Exercises

4.1   If factor C is assigned to column 4 of an L_8 OA and factor B assigned to column 6, which column will estimate the interaction between them?

4.2   Design an experiment to study five factor effects, A, B, C, D and E, and two interactions; $A \times D$ and $A \times C$. Select the suitable OA for the experiment and identify the columns for the interactions.

4.3   In a certain robotic welding process, the objective of the experiment was to increase the weld strength of a bracket welded to steel pipes and thereby reduce scrap rate. The Taguchi team had identified six main effects: A, B, C, D, E and F. Two interactions ($B \times C$ and $C \times E$) were identified to be most likely to interact and therefore of great interest to the team. Which OA would you recommend to the team?

4.4   For exercise 3.9, how would you construct a problem graph and a modified problem graph for the robotic welding experiment?

4.5   An experimenter wishes to study the following two-level factors and two-factor interactions:

Factors: A, B, C, D, E, F, G and H.

Two-factor interactions: A × B, A × C, A × D, A × E, F × G and F × H.

Construct a problem graph and a modified problem graph for the experiment.

4.6 The Environmental Protection Agency has identified four factors, A, B, C and D, each at two levels, that are significant in their effect on the air pollution level at a photographic film production facility. The agency feels that the interactions A × C and B × C are important. Show an experimental design using an OA that can estimate the above effects.

4.7 A baseball manager strongly believes that five factors, each at two levels (A, B, C, D and E) are significant in affecting runs batted in. The manager also believes that the interactions B × C and B × E are important and therefore to be studied. Show an experimental design using an OA that can estimate the above effects.

References

1. Lee, N.S., Phadke, M.S., and Keny, R. (1989) "An Expert System for Experimental Design in Off-line Quality Control", Expert Systems, Vol. 69, No. 4, pp. 238–249.
2. Antony, J. and Kaye, M. (1996) "An Application of Taguchi's Robust Parameter Design Methodology for Process Improvement", Journal of Quality World Technical Supplement, March, pp. 35–41.

5 CLASSIFICATION OF FACTORS AND CHOICE OF QUALITY CHARACTERISTICS

5.1 Classification of factors in Taguchi's experimental design methodology

For manufacturing process optimization problems using Taguchi methods, the following factors are of interest to the experimenters:

- control factors;
- noise factors;
- signal factors.

A block diagram as shown in Figure 5.1 depicts those factors that influence the response (or the quality characteristic) of a product or process. In the block diagram, y stands for the response (quality characteristic/output). Here we consider only the case of a single response as the extension to multiple responses is straightforward.

5.1.1 Control factors (x)

Control factors are those factors that can be controlled easily during actual production conditions or standard conditions. For example, in an injection-

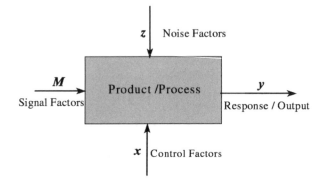

Figure 5.1.   Block Diagram of a Product/Process

moulding process, mould temperature, barrel temperature, injection speed, injection pressure, cure time, etc. are control factors. These are also called design parameters, the levels of which are determined by the design engineers. It is the objective of the design activity to determine the best levels of these factors to achieve the product or process robustness. Here *robustness* refers to making products or/and processes insensitive to various sources of variation.

Although control factors are studied to establish their ideal values to accomplish the objective of the experiment, it would be useful to classify the effects of control factors on the quality characteristic (or response), as follows:

- control factors affecting the mean quality characteristic (or response) only;
- control factors affecting the variability in response only;
- control factors affecting the mean response and response variability; or
- control factors affecting neither the mean response nor the response variability.

Control factors affecting the mean quality characteristic (or response) only. A control factor that affects only the mean response is called an *adjustment factor*. An adjustment factor in Taguchi style of experimentation is used to bring (or adjust) the mean response of a process or product onto the target value (Figure 5.2a).

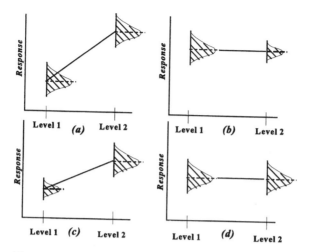

Figure 5.2.   Types of Control Factors

Control factors affecting the variability in response only. A control factor that affects the response variability but not the mean response can be used to reduce the variation of a process or product (Figure 5.2b).

Control factors affecting the mean response and response variability. A factor that influences both the mean response and response variability must be handled and used very carefully. However, such a factor allows some flexibility in balancing target requirements (Figure 5.2c).

Control factors affecting neither the mean response nor the response variability. A factor that has no impact on the mean response or response variability is not a useless factor, the choice of level of such factors depending advantageously on various constraints such as cost, convenience and so on (Figure 5.2d).

In industrial experimentation, it is not often that a factor influences the mean response only or the response variability only. Generally, it is more likely that a factor that influences the mean response also influences the response variability to some extent, and vice versa.

5.1.2 Noise factors (z)

Noise factors are those factors that are difficult, expensive or hard to control during actual production conditions but may be controllable during experimentations [1]. Manufacturers have no direct control over noise factors and these vary with the customer's environment and usage. These factors cause the performance characteristic of a product to deviate from its target or nominal value. The levels of the noise factors change from one unit to another, from one environmental condition to another and from time to time. Only the statistical characteristics such as the mean and variance of noise factors can be known or specified, the actual values in specific situations cannot be known. In an injection-moulding process, ambient temperature, relative humidity, operator, machine ageing, etc. are generally deemed as noise factors. Now, one might say that the noise factors such as ambient temperature or humidity could be controlled, which is true. However, if one considered the cost of controlling this noise in a large manufacturing environment, it is easy to see how prohibitive it might become from a monetary point of view. These noise factors are the sources of variation in products and processes and therefore the cause of poor quality. The output response deviates from its target performance because of the existence of these undesirable noise factors. Taguchi advocates the use of ***robust design*** to identify the control factor settings that will dampen the effect of the noise factors and thus reduce the variation of response from its target performance at low cost. Figure 5.3 summarizes the objective of the robust design method. The robust design method would increase

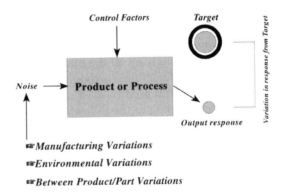

Figure 5.3. Objective of Robust Design

process yields, increase product life times, reduce scrap rates, increase product reliabilities and improve product performance.

Consider an injection-moulding process used for making plastic components. The output of interest to the experimenter is the shrinkage measured in millimetres. The target value for the shrinkage is zero. The average shrinkage obtained from a sample of 50 taken from the normal production settings was about 1.6 mm. A robust design was carried out with the aim of minimizing the effect of noise (i.e. ambient temperature, percent regrind, relative humidity, etc.) so that the average shrinkage can be brought as close as possible to the target. The optimal settings of control factors have produced an average shrinkage of 0.23 mm. Therefore the objective of a robust design is to determine the best settings for control factors which will make the process output insensitive to all sources of noise.

The noise factors can be classified into three categories (Figure 5.4):

- *inner noise factors;*
- *outer noise factors; and*
- *between-product (or product-to-product) noise factors.*

Inner noise factors (or deterioration).   Inner noise factors are those that cause the variations in performance that are internal to the product or process design. Tool wear, machinery ageing, oxidation (or rust), deterioration of parts or components, fade of colour, etc. are examples of inner noise factors.

Outer noise factors.   Outer noise factors are those that affect the performance of a product or process from an external standpoint. Environmental factors such as ambient temperature, humidity, pressure, operators, dust, supply voltage, thermal shock, vibration, etc. are examples of outer noise factors.

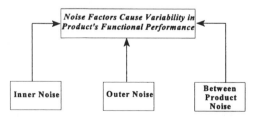

Figure 5.4.   Classification of Noise Factors

Between-product noise factors. Between-product noise factors relate to the variability in performance seen in products made under exactly the same manufacturing environment. For example, a motorbike giving 20 miles/litre of fuel and another of the same brand giving only 17 miles/litre. This is also called unit-to-unit noise, which is the difference in product's performances due to variability in raw materials, manufacturing equipment and assembly process.

Consider the development and production of a braking system. An outer noise that would affect the performance of this system might be the condition of the road surface—wet or dry. An example of inner noise might be related to pad wear. In this example, between-product noise is the variation in performance observed between the braking systems produced in the same manufacturing environment, made by exactly the same method.

It is not important to classify accurately the type of noise factors affecting the performance of a product or process. It is, however, important that engineers design products and processes to be robust (or insensitive) to all types of noise in the user's environment. This means that the goal of an engineer is to search for and achieve 'robust' designs for products and processes [2]. It is only through achieving this state that engineers can enhance their technological capabilities.

5.1.3 Signal factors (M)

These are the factors that are set by the user or operator to achieve the target performance [3]. Generally, signal factors have a significant impact on the process mean and therefore can be used to bring the mean onto the target value. For example, the steering angle for the steering mechanism of an automobile which determines the turning radius is a signal factor. The speed-control setting on a table fan is another example of a signal factor for specifying the amount of breeze. Other examples of signal factors are the 0 and 1 bits transmitted in a digital communication system and the original document to be photocopied by a photocopying machine. The signal factors are selected by the design engineers, based on the engineering knowledge of the product being developed. A response (or quality characteristic) in the design of an experiment in which the signal factor takes a constant value (in which case it is not included as a factor) is called a *static characteristic*. When the signal factor takes a number of values, the quality characteristic is then called a ***dynamic characteristic***. In an injection-moulding process, the signal factor can be the dimension of the die. Similarly, in a fire detector mechanism, the signal factor can be the smoke temperature. Taguchi accentuates that all manufacturing systems should be treated as dynamic and not static for achieving the greatest optimization [4].

5.2 The role and contribution of noise factors in industrial experiments

In the past, many engineers tried to deal with problems related to noise factors by controlling the noise factors themselves: they hermetically sealed components sensitive to humidity; isolated components sensitive to vibration; air conditioned operations in which temperature was a key factor, and so on. Taguchi proposed that such control actions to remove or minimize the effects of noise be used only as a last resort, since they are quite expensive to control for experimentation purposes. Instead he advocated determining the best levels for control factors (or design parameters) which will dampen the effect of noise on product and process performance. To accomplish this objective, Taguchi uses *parameter design*. Parameter design is used to improve quality without controlling or removing the cause of variation, to make the products and processes robust against noise factors. Taguchi refers to the situation in which a product or process exhibits consistent, high-level performance, despite being subjected to a wide range of customer and manufacturing conditions, as the state of robustness. Robustness is synonymous with high quality and high reliability. Industrial engineers are often interested in studying the role and effect of these noise factors which cause excessive variability in the functional performance of products. In parameter design (also called robust design), it is essential to study the effect of noise factors on the functional performance characteristic of a product or process. Noise factors are responsible for causing a product's functional performance characteristic to deviate from its target value. This deviation of the functional performance characteristic of a product causes a loss to society. This quality loss can be reduced by obtaining the optimal conditions for control factors or design parameters. In other words, the determination of the optimal settings of control factors can minimize the deviation of the performance characteristic of a product from its target and thereby minimize the quality loss to the society. Figure 5.5 illustrates the role and contribution of noise factors in industrial experiments.

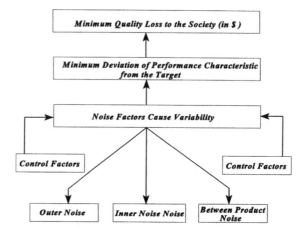

Figure 5.5.   Role and Contribution of Noise Factors

5.3   Design for robustness—the key to improve product and process quality

Robustness is defined in terms of the performance of the process or product and how noise factors affect that performance in real-life situations [5]. Robust processes are those that are insensitive to the materials, environmental and variation in machine parameters; whereas robust products are those that are insensitive to the manufacturing process, usage conditions, materials and so on. Creating a robust process or product is like putting shock absorbers on a car. Shock absorbers enable the driver to get a smooth ride irrespective of the road conditions, which are not controllable.

But why is robustness important? Using robustness, variation in the functional performance of both products and processes can be reduced without increasing costs. Trade-offs exist between the degree of robustness and costs. Too much robustness might add unnecessary cost. Nevertheless, achieving robustness is critical to minimizing quality loss. It is important to note that robustness reduces loss to society, which includes customer dissatisfaction, loss of market share, bad reputation and so on.

Creating robustness is very similar to potential failure mode effect and criticality analysis (FMECA) used in product and process development. Anil Rastogi [6] illustrates a real industrial application of FMECA to achieve robust design and improve product reliability of an infusion pump. FMECA in this context is useful to identify what might go wrong and the appropriate design changes. If robustness is achieved for designs (product or process) upstream, then fewer problems occur in manufacturing and in the field; fewer changes are required; costs associated with scrap, repair and rework are reduced; warranty claims are reduced and customer satisfaction is enhanced.

5.4 Treating noise factors incorrectly

It is very important to note that unless noise is deliberately induced into the experiment, there will be no assurance for the achievement of product or process robustness. A common mistake is to include noise factors in the OA for control factors. For example, an experimenter wishes to study four factors; A, B, C, and D, using an L_8 OA. Each factor needs to be evaluated at two levels. The experimenter also wants to study a noise factor, N, and two interactions: AC and BC. The experimental layout chosen by the experimenter using the L_8 OA is shown in Table 5.1.

The analysis of this experiment will provide the following information for the experimenter:

- information on how the control factors (or process parameters) affect the response;
- information on how the noise factor, N, affects the response.

Because the noise factor was included in the L_8 array for the control factors, the experiment cannot provide any information whatsoever about the sensitivity of the control factors to the noise. Therefore the experimenter may fail to identify the control factor settings for achieving process robustness.

Table 5.1. Experimental Layout for the L_8 OA

Run	A	B	D = AB	C	AC	BC	N	Response 'y'
1	1	1	1	1	1	1	1	y_1
2	1	1	1	2	2	2	2	y_2
3	1	2	2	1	1	2	2	y_3
4	1	2	2	2	2	1	1	y_4
5	2	1	2	1	2	1	2	y_5
6	2	1	2	2	1	2	1	y_6
7	2	2	1	1	2	2	1	y_7
8	2	2	1	2	1	1	2	y_8

In such circumstances, it will be wise to use Taguchi's approach for design of experiments whereby we assign control factors and noise factors in the control and noise array of the experimental layout. The idea of using such an approach is to determine the best combination of control factor settings that will minimize the effect of noise and thereby make the process robust against uncontrolled sources of variation.

5.5 Taguchi's product array approach to experimentation

Taguchi's product array approach consists of two parts: a control array (or inner array) and a noise array (or outer array). A **control array** is a matrix with rows and columns. The columns of the matrix represent the control factors (or design parameters) and rows represent the settings of the control factors to be used in the experiment. The **noise array** is used for measuring the effects of noise factors. This array specifies how the noise factors will be varied during the experiment. The columns of a noise matrix represent the noise factors, and the rows of the matrix represent different combinations of the levels of noise factors to be used in the experiment. Figure 5.6 illustrates a simple 'product array' (inner array × outer array) experiment with three control factors at two levels in the control array and two noise factors at two levels in the noise array.

If the control array has m rows and the noise array has n rows, then

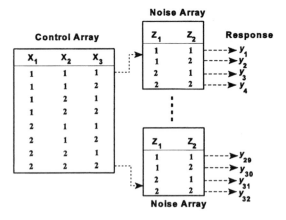

Figure 5.6. Taguchi's Product Array Experiment Format

the total number of experimental runs in the product array layout is $m \times n$. For each of the m rows of the control array, the n rows of the noise array provide n or more repeat observations on the performance characteristic (or response). In Figure 5.6, each of the eight runs in the control array should run against each of the four in the noise array. Hence the experiment consists of a total of 32 experimental runs. It is good practice to examine the interaction between control and noise factors to achieve the state of robustness.

The repeat observations on the response from each test run in the control array are then used to compute a performance statistic called the *signal-to-noise ratio*. Dr Taguchi developed the concept of signal-to-noise ratio (SNR) in engineering to evaluate the performance of a system. This concept originated in the field of electrical engineering. The objective of using SNR is to develop systems which are robust against noise factors [7]. The SNR measures the sensitivity of the quality characteristic (or response) being investigated in a controlled manner, to those external undesirable disturbances (or noise factors) which are uncontrollable or expensive to control. Taguchi effectively applied the concept of SNR to establish the optimum condition from the industrial experiments. The goal of any industrial experiment is to determine the combination of process parameter (or control factor) settings that will maximize the SNR. The larger the SNR, the less variability around the mean and the smaller the quality loss. A high value of SNR connotes that the signal is much higher than the random effects of the noise factors. A process operation consistent with high SNR always yields the optimum quality with minimum variance. A detailed explanation of SNR can be found in Chapter 9.

5.6 Choice of quality characteristics for industrial experiments

A quality characteristic (or response of interest) can be defined as a characteristic that an experimenter wants to measure in an experiment, in order to evaluate the quality of a product. For an industrial experiment, selecting an appropriate quality characteristic is important so that the characteristic is both additive and monotonic [8]. A quality characteristic with good additivity property is mandatory for efficient, reliable and reproducible experimentation. The selection of quality characteristics requires an engineer to draw upon his/her engineering knowledge of the process or product under study. Taguchi emphasizes that the quality characteristic (or response) should be chosen such that the effects of control factors are additive. Additivity means no interaction between the factors. In other words, the total effect of factors on the output quality characteristic (or performance characteristic) is the sum of all main effects. However, Kacker [9] pointed out that it is unrealistic to assume that one can always find relevant performance characteristics that are additive in the effects of control factors.

In designing a product, we are usually interested in improving the product reliability by increasing the life of the product. In designing a manufacturing process, we are usually interested in maximizing the yield of the process by reducing the scrap rate or number of defects. The final success of the process or the product depends on how well responses (i.e. reliability, yield, etc.) meet the customer's expectations. However, such responses are not necessarily suitable as quality characteristics for optimizing process or product design. The following guidelines may be useful to engineers in industry in selecting the quality characteristics for industrial experiments.

- Identify the ideal function or the ideal input-output relationship for the product or the process. The quality characteristic should be related directly to the energy transfer associated with the basic mechanism of the product or the process.
- Quality characteristics should, as far as possible, be continuous variables.
- The quality characteristic should be monotonic, at least in the range of the experiment. That is, the effect of each process variable (or control factor) on robustness should be in a consistent direction, even when the settings of other process variables (or control factors) are changed. In fact, it is difficult to judge the monotonicity of a quality characteristic before conducting experiments. In such situations, OA experiments followed by confirmatory experiments are the only way of determining whether the quality characteristics have monotonicity.
- Try to use quality characteristics that are easy to measure. The availability of appropriate measurement techniques is often an important consideration in the selection of a good quality characteristic.
- It is important to ensure that quality characteristics are complete. That is, they should cover all dimensions of the ideal function or the input–output relationship.

The concepts behind these guidelines are rather new. They will become clear through the examples presented in the following sections. Finding quality characteristics that meet all of these guidelines is sometimes difficult or simply not possible.

5.6.1 Examples of quality characteristics or responses

In this section, we will look at some examples of quality characteristics, analyse how well they meet the guidelines of the previous section and examine the impact on experimental efficiency.

Example 1 Coating thickness as a quality characteristic. A certain coating process results in various problems such as poor appearance, low yield, orange peel and voids. Too often, experimenters measure these characteristics as data and try to optimize the response This is not sound engineering, because these are simply the symptoms of poor function. It is not the function of the coating process to produce an orange peel. The real problem is the variability of the coating process due to noise factors such as variability in viscosity, ambient temperature and so on. We should measure data that relate to the function itself and not the symptom of variability.

One fairly good characteristic to measure for the coating process is the coating thickness. The function of the coating process is to form the coating layer. Symptoms such as orange peel and poor appearance result from excessive variability of coating thickness from its target. It is sound engineering strategy to measure the coating thickness and to determine the combination of control factor settings that will minimize the coating thickness variability around its target value.

Example 2 Spray-painting process. This example clearly illustrates the importance of energy transfer in selecting a quality characteristic. In a spray-painting process, sag is considered to be a common defect. It is caused by formation of large paintdrops that flow downward due to gravity. Is the distance through which the paintdrops sag a good quality characteristic? The answer is no, because the distance through which the paintdrops sag is basically controlled by gravity and it is not related to the basic energy transfer in spray painting. However, the size of the drops created by spray painting is directly related to energy transfer and thus is a better quality characteristic for the spray-painting process.

Example 3 Body–door alignment of an automobile. In this example, we consider the alignment of a body and a door of an automobile. Assume

a and b are the door and body dimensions of the automobile, respectively. The function $y = b - a$ is a poor quality characteristic. Here variability arises from the variability of a and b, the stamping processes of door and body components. If y is chosen as the quality characteristic to be measured, then the measurements of y may be full of interactions. The real engineering problem is the variability in the stamping process itself. The choice of y as a quality characteristic has a poor chance of being reproducible downstream. It would be better to treat a and b as completely independent quality characteristics to be optimized. This would improve the reproducibility of results for the body–door alignment of the automobile.

Example 4 Paper jamming in a photocopying operation. Paper jamming is an important customer-observable problem in photocopier operation. Is the number of jams per thousand sheets of paper a good quality characteristic? The two main problems that arise in a paper feeding mechanism are: no sheet fed or multiple sheets fed. Therefore, a quality characteristic that measures the force needed to pick up one sheet (to avoid paper jamming) rather than two or more sheets of paper would be a better quality characteristic.

5.6.2 *Multiple quality characteristics or responses*

In the above section, we have been discussing the selection of one quality characteristic (or response) for an industrial experiment. Often, we need to consider more than one quality characteristic of a process or product. In such cases, each response must be analysed separately from the available data. An overall selection of factors that minimize the sacrifice of quality characteristics may require a trade-off of factor levels. Hence it may be important to review the selection of factor levels.

It is important to note that sometimes an improvement of one quality characteristic may result in the degradation of another. In order to attenuate such effects, it is important to identify a factor that affects one quality characteristic but not another. This strategy becomes important while optimizing conflicting characteristics such as power and efficiency of an automobile engine. By including many factors in the experiment, the likelihood of identifying factors that can be used selectively to improve both quality characteristics simultaneously may be increased.

Table 5.2. Quality Characteristics for a Variety of Manufacturing Processes

Problem number	Type of process	Nature of the problem/objective of the experiment	Appropriate quality characteristic (or response)
1	Die casting process	To increase the hardness of a die cast engine component	Rockwell Hardness
2	Wave-soldering process	To reduce the average defective rate of solder joints on the Printed Circuit Boards	Number of defective joints
3	Injection moulding process	To reduce parts shrinkage	Percent shrinkage
4	Thermoplastic tape winding process	To study the impact of process parameters on interply bond strength	Bond strength
5	Biscuit baking process	To reduce variability in biscuit length and weight	Length, Weight
6	Lathe Turning operation (machining process)	To study the effects of cutting parameters on surface finish	Surface roughness
7	Wire bonding process	To reduce the defect rate from broken wires	Wire pull strength
8	Coil spring manufacturing process	To reduce variability in the tension of coil springs	Spring tension
9	Extrusion process	To reduce the post-extrusion shrinkage of a speedometer cable casing	Post-extrusion shrinkage
10	Gold plating process	To reduce variation in gold plating thickness and then bring the mean thickness to the target value	Plating thickness

Table 5.2. *Continued*

Problem number	Type of process	Nature of the problem/objective of the experiment	Appropriate quality characteristic (or response)
11	Hot-plastic forming process	To reduce the rework rate of a diesel injector on the assembly and test operation	Pull-out load
12	Injection moulding process	To determine the process parameter settings to produce a product with the proper dimensions	Length and Width
13	Cylindrical lapping process	To determine the effects of process parameters on surface roughness	Surface roughness
14	Casting process	To study the critical variables that have impact on the quality of castings	Surface roughness
15	Hydraulic hose wire curl forming process	To study the effect of die parameters on curl formation	Curl length
16	Deburring operation	To improve the efficiency of deburring operation	Material removal rate
17	Metal Inert Gas (MIG) welding process	Unreliable welded joints and thereby high scrap rate	Strength of the weld
18	Screen Printing process of thick film hybrids	To minimise variability in print thickness	Print thickness
19	Shock absorber optimisation process	To reduce variability in damping forces of a shock absorber	Extension force
20	Chemical wire bonding process	Low process capability and hence high rework costs	Bond strength

Table 5.2.   *Continued*

Problem number	Type of process	Nature of the problem/objective of the experiment	Appropriate quality characteristic (or response)
21	Die cast machine process	To reduce or eliminate the cold shot (a defect related to die cast process)	Surface area of cold shot
22	Reaction Injection Moulding Process	To reduce porosity (surface defect)	Porosity
23	Manufacturing process of an industrial thermostat	To improve the performance (i.e., reliability) of the thermostat	Life
24	Assembly process of a starter relay	To reduce variability in dropout voltage and circuit voltage drop	Dropout voltage, Circuit voltage drop
25	Hot plastic forming process	To maximise the failure load of the retaining ring	Pull-out force
26	Resistance Welding Process	To obtain maximum welded joint resistance	Welded joint resistance
27	Wire bonding process	Process optimisation	Pull strength
28	TV picture tube manufacturing process	To reduce performance variation of electron guns	Cut-off voltage
29	Surface mount process	To reduce scrap rate and improve field reliability	Shear strength, Adhesive deposit diameter
30	Foundary operation	To reduce shrinkage porosity	Porosity

5.6.3   Quality characteristics for industrial experiments

It is important to separate the objective of the experiment from the quality characteristic to be measured. Some careful thought in selecting the response is repaid many times over during the experiment [10]. Selecting the wrong thing to measure can make the experimental process immaterial, in the sense that the conclusions from the experiment are obvious. The choice of the right quality characteristic is essential for the success of any industrial experiment. In order to assist industrial engineers to select an appropriate quality characteristic (or response), Table 5.2 has been developed, which covers a variety of manufacturing process problems and the suitable response of interest to experimenters for each associated process. Table 5.2 was generated as a result of a thorough investigation of various industrial case studies and the authors' expertise and skills in the area of study.

Exercises

5.1   Illustrate the effect of control factors on the performance characteristic of a process.

5.2   What is robust design?

5.3   Explain the role and contribution of noise factors in industrial experiments.

5.4   Explain why robustness is important for achieving high-quality products.

5.5   Discuss the importance of selecting good quality characteristics.

5.6   Is yield a good quality characteristic? What are the key points to be considered while selecting quality characteristic(s) for an industrial experiment?

References

1. Tsui, K-L. (1992) "An Overview of Taguchi Method and Newly Developed Statistical Methods for Robust Design", IEE Transactions, Vol. 24, No. 5, November, pp. 44–55.
2. Pal, B.K. et al. (1995) "One Day Programme on Quality Engineering", Course Manual, SQC and OR Division, ISI Bangalore, India, 21 April.
3. Taguchi, G. and Phadke, M.S. (1984) "Quality Engineering through Design Optimisation", IEEE, pp. 1106–1113.

4. Taguchi, G. (1993) "Taguchi on Robust Technology Development", ASME Press.

5. Snee, R.D. (1993) "Creating Robust Work Processes", Quality Progress, February, pp. 37–47.

6. Rastogi, A.K. and Sahni, A.K. (1995) "FMECA—A Tool to Achieve Robust designs", 5th World Quality Congress on Total quality, pp. 149–152.

7. Kapur, K.C. and Chen, G. (1988) "Signal-to-Noise Ratio Development for Quality Engineering", Quality and Reliability Engineering International, Vol. 4, pp. 133–141.

8. Phadke, M.S. and Taguchi, G. (1987) "Selection of Quality Characteristics and S/N ratios for Robust Design", IEEE, pp. 1002–1007.

9. Kacker, R.N. (1992) "Taguchi's Parameter Design: A Panel Discussion", Technometrics, Vol. 34, No. 2, p. 139.

10. Grove, D.M. and Davis, T.P. (1992) "Engineering Quality and Experimental Design", Longman Scientific and Technical.

6 A STRATEGIC METHODOLOGY FOR TAGUCHI DESIGN OF EXPERIMENTS

6.1 Introduction

Design of experiments and Taguchi methods are well-established methodologies in which only statisticians are formally trained. Design of experiments (DOE) is a powerful approach for investigating the optimal combination of process parameters in respect of quality performance. Taguchi methods (TMs), on the other hand, are used for maximizing product and process robustness by reducing variation due to undesirable external disturbances which cannot be controlled during actual production conditions. Over the past decade, both DOE and TMs have had unprecedented success in showing how statistical methods assist organizations in manufacturing high-quality products at low costs. However, recent research has shown that the application of such statistical methods by the engineering fraternity in manufacturing companies is limited due to lack of skills required in manufacturing and inadequate statistical knowledge for problem solving. In order to bridge this gap, this chapter presents a practical and strategic methodology for Taguchi methods, to tackle and solve manufacturing process quality problems in manufacturing companies. The methodology is easy to understand and readily accessible to the engineering fraternity for solving quality problems in real-life situations. The objective of this step-by-step approach is to assist industrial engineers with limited statistical knowledge to tackle process quality problems using TMs.

6.2 Devised methodology

The aims of the methodology are:

- To assist industrial engineers with limited statistical knowledge to apply Taguchi methods effectively for solving manufacturing process quality problems.
- To assist manufacturing companies to tackle and solve manufacturing process problems in a more systematic and organized manner in order to augment their quality control performance.
- To assist with the encouragement of the wider use of Taguchi methods (TMs) as a powerful problem-solving tool in manufacturing companies.

The devised methodology focuses on a generic method of classifying manufacturing process problems in terms of product quality and process effectiveness and shows how it assists companies to tackle and confront manufacturing quality problems more efficiently. The developed methodology pays attention to modern statistical analysis methods and interpretation by using rule-based methods and graphical tools for easy and rapid understanding of the results from statistical analysis. The devised methodology consists of 10 steps, and is demonstrated by the flowchart shown in Figure 6.1.

Step 1   General overview of the product or process.   Studies and several brainstorming sessions showed that a general understanding of the process (or product) to be investigated is highly recommended for identifying areas where quality problems are a fairly common occurrence. The methodology suggests the use of flowcharts for studying the various operations associated with the relevant process and for identifying areas where problems can occur. Flowcharting the relevant process is essential; however, many experimenters often skip this invaluable tool as they feel it takes too long and they already understand their processes. Unfortunately, they are only partially correct. They understand the processes but only from their own limited perspective [1]. By constructing a flowchart, all team members

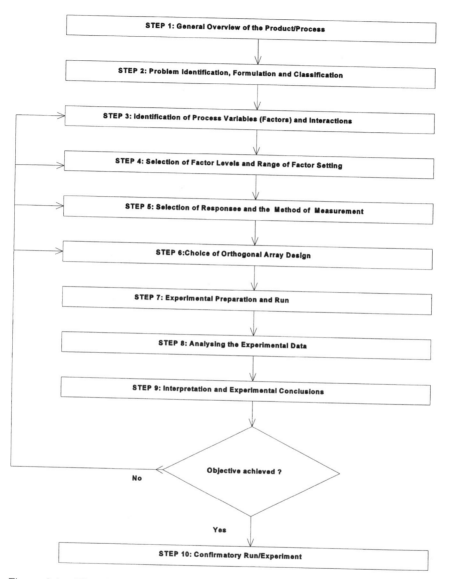

Figure 6.1.   Flowchart of the Devised Methodology

involved in the experiment come to a common understanding of the process under investigation.

Having used the flowcharts, it is good practice to then use process failure mode and effect analysis [2, 3] for prioritizing possible problems in the order of their severity. It is important to note that not all process quality problems can be tackled by the application of Taguchi methods. However, to investigate the impact of process parameters or factors on the output performance characteristic (or response) or to make the process robust against unknown sources of variation in the user's environment, Taguchi methods are often an invaluable tool.

As part of the initial investigation of the process, it is recommended to study various machines, materials, measurement systems and so forth used for the process to be investigated. This can be accomplished by building a multidisciplinary team of people who can contribute knowledge from different perspectives. The team usually includes members from process engineering, production, quality engineering and machine operators.

Step 2   Problem identification, formulation and classification.   Having studied the relevant process using process flowcharts and process failure mode and effect analysis, the user (i.e. an industrial engineer) is asked to identify, formulate and classify the engineering problem from which an appropriate analysis can be made. The methodology suggests the use of brainstorming for identifying the problem. Problem identification is very critical for any industrial experiment as the experimental and analysis part is based on this. Hence it is necessary to have a sound knowledge of the process to be investigated. Some of the generic process problems in manufacturing where Taguchi methods can be utilized are:

- *customer feedback*, indicating that the process performance needs to be improved;
- *excessive variability* in processes, leading to high scrap rates;
- *incapable processes*, resulting in a high rate of rejected parts;
- *low process yields* as the process is not operating at the optimum condition.

Brainstorming is a very important stage of the whole Taguchi experimentation process. It is an activity that promotes team participation, encourages creative thinking and generates many ideas in a short period of time. The time needed for a brainstorming session is mostly dependent on the nature of the problem and the number of people involved in the team. Taguchi considers brainstorming as an essential step in the design of effective experiments [4]. Taguchi does not institute a standard method for brain-storming as applicable to all situations in manufacturing companies. If the brainstorming fails, there is little chance of salvaging the experiment and any further efforts could be a waste of resources.

During brainstorming, all the ideas must be examined critically and nar-rowed down. Brainstorming groups should be small, i.e. 4–8 people. Ideas should not be challenged during the brainstorming session. When the brain-storming is complete, the list can be critiqued and reduced to a workable number of ideas. The guidelines for brainstorming can be obtained from Barker's text, *Quality by Experimental Design* [5].

Having identified the problem to be studied, the nature of the problem has to be formulated and the possible causes (or influential factors) of the problem must be thoroughly investigated using cause and effect analysis (CEA). CEA is a tool for identifying the most probable causes affecting a problem. It can help in the analysis of cause and effect relationships, and it can be used iteratively in conjunction with brainstorming [6]. The major cat-egories of causes are methods, materials, machines, people, environment and measurement. Pareto analysis can then be used to separate out the main causes of the problem.

Step 3   Identification of process parameters (or factors) and interactions of interest.   Once the problem is understood, the next step is to identify process parameters (or factors) associated with the problem. It is very important to assure that the selected process parameters really provide the necessary and pertinent information about the problem. Some possible ways of identifying potential factors are through the use of brainstorming, historical data and the use of engineering knowledge of the process [7, 8]. As the team brainstorms the problem (in step 2) to explore the possible causes, many factors are generated and listed in the cause and effect diagram.

How many factors should be included in the study? This commonly asked question is related to the size of the experiment and raises the exper-imental cost and the time needed. In the early stages of investigation, an orthogonal array (OA) can be used to screen many factors. Therefore, as many factors as possible should be selected within the stated experimental cost and time constraints [9].

Having identified the factors to be studied, the next stage is to classify them into control and noise factors. The selection of signal factors (if any) depends on the nature of the problem and the degree of optimization required for the experiment. As the methodology is developed for static quality characteristics, importance is imparted to control and noise factors.

Once the factors are identified, the team has to list the potential inter-actions (among the control factors) of interest for the study. The next questions are "Should an interaction be replaced by an additional factor?" and "Do we need to study the interactions at all in the first place?" The team must be prepared to answer these questions prior to choosing an appropriate OA for the experiment. If noise factors are considered for the experiment, main effects of noise factors should have paramount importance compared with the interaction effects among the noise factors. In order to minimize the size of the experiment, interactions among the noise factors can be ignored and, generally, exploring such interactions are just a waste of resources. Instead, it is wise to explore the interaction between control and noise factors for achieving robustness.

Step 4 Selection of factor levels and range of factor setting. Determining the levels of selected factors from brainstorming is another major concern to many experimenters in industries. For initial screening experiments, the number of levels of each factor should be kept low, two levels if possible. If it is expected that the relationship between the input and output is linear, then two levels will be recommended. If a non-linear function is expected, then three or more levels should be used to quantify the non-linear (or curvature) effect in the response.

For quantitative factors such as temperature, pressure, speed, power and so on; generally two levels (i.e. low and high) are recommended in the early stages of investigation and the levels chosen should cover the entire range of interest. For qualitative factors, such as type of catalyst, type of machine, type of material and so on, generally more than two levels are required.

Having determined the number of levels required for each factor, it is advised to specify the range of operation for each control factor and noise factor (if any). It is usually best to experiment with the largest range feasible, so that the variation inherent in the process does not mask the factor effects on the response. If the range of a factor setting is wrong (i.e. too low or high), then one may not be able to observe the effect of that factor on the response. Choosing the appropriate levels for factors is not an easy task. It is advisable to select factor levels far enough apart so that, if there is an important effect, it will be observed; but not so far apart that a different process is operating. For example, consider an experiment to determine the optimum temperature for a process involving heated water: 220°F and 225°F are too close because experimental error will eliminate any potential effect; 150°F and 300°F are too far apart because one is below the boiling point of water and the other is above, indicating that a different process is probably in action.

Step 5   Selection of response (or quality characteristic) and the method of measurement.   Selecting the appropriate response (or quality characteristic) is critical to good experimentation. For example, continuous responses will typically provide more information and require fewer resources than categorial type (or discrete) responses. To identify a good response, it is suggested to start with the engineering or economic goal. Having determined the goal, identify the fundamental mechanisms and physical laws affecting this goal. Finally, choose the response to increase the understanding of these mechanisms and physical laws [10]. The response should be related as closely as possible to the basic engineering mechanism of the technology. For example, suppose a soap manufacturer faces an underweight problem that a certain proportion of the bars of soap weighs less than the stated weight marked on the label. At first, it might seem that the best response for studying this problem is the weight of the bar. This response, however, would not provide a good understanding of the basic physical mechanisms involved in controlling the weight of the bar during production. It is important to note that the weight of the bar is a product of density and size. For a particular type of soap, density is controlled by the amount of air mixed into the soap. Similarly, the bar size is controlled by the x–y–z settings of the tool used to cut soap slabs into bars. In this particular example, the engineering or economic goal is to eliminate the underweight problem. The physical mechanisms affecting this goal are the mixing and cutting processes. The response that increases the understanding of these physical mechanisms is density and x–y–z dimensions of the soap bars.

The following points are useful while selecting the quality characteristic or response for an industrial designed experiment:

- use quality characteristics (or responses) which can be measured precisely, accurately and with stability;
- use continuous responses where possible;
- for complex or intricate systems (or processes), select responses at the sub-system level and perform experiments at this level prior to attempting overall system optimization;
- use quality characteristics which are practical, in the sense that they are easy to measure;
- it is good practice to check the precision and accuracy of the measurement system when measuring the response of interest.

The method of measurement of the response (or quality characteristic) should be well defined. This includes choosing the measurement and processing equipment to be used, how to measure, where to measure and where to document the data. If the capability of the measurement system has not already been evaluated, it should be evaluated before the experiment is run. The gauge repeatability and reproducibility (GRR) study can be used to study the major sources of variation such as operator reproducibility and equipment (or gauge) repeatability [11].

Having identified the quality characteristic and its method of measurement, the next stage is to identify the type of quality characteristic that needs to be optimized. This book is focused on four types of quality characteristics, such as smaller-the-better (STB), larger-the-better (LTB), nominal-the-best (NTB) and classified attributes (CA). The selection of any of these characteristics is dependent on the objective/goal of the experiment.

- *Smaller-the-better (STB) quality characteristics:* this type of quality characteristic is considered when measuring the porosity, vibration, fuel consumption of an automobile, wear of a bearing and shrinkage of a plastic component.
- *Larger-the-better (LTB) quality characteristics:* this type of characteristic is considered when measuring strength, efficiency, life and the mileage of an automotive per gallon of fuel.
- *Nominal-the-best (NTB) quality characteristics:* for this type of characteristic, one may consider measuring dimensions such as diameter, thickness and width. Other examples include force, pressure, viscosity, area and volume. A nominal (or target) value is always specified for this characteristic and minimal variability around the target is desired.
- *Classified attribute (CA) quality Characteristics:* examples of classified attribute characteristics include good/bad and grade A/B/C/D. For example, in a certain welding process, the severity of crack can be classified as "no crack", "mild crack", "moderate crack" and "severe crack".

Step 6 Choice of orthogonal array (OA) design. Orthogonal arrays allow experimenters to estimate both the main effects and their interaction effects in a minimum number of experimental runs. The choice of OA is very critical as it depends on a number of factors, including the total degrees of freedom required for the main effects and their interactions, resolution of the design, objective of the experiment and, of course, cost and time constraints.

The orthogonal arrays are selected for inner arrays (i.e. for control factors) and outer arrays (if noise factors are to be studied). Linear graphs and interaction tables are used to assign factors and two-factor interactions in appropriate columns of the OA (assuming that three and higher-order interactions are negligible).

To assure that the chosen OA design will provide sufficient degrees of freedom for the contemplated experiment, the following inequality must be fulfilled:

$$V_{OA} \geq V_{\text{required for main and interaction effects}}$$

Having selected the appropriate OA for the experiment, the next step is to assign factors and interactions of interest to the suitable columns of the array. Problem graphs and confounding patterns must be constructed while assigning factors and interactions to the columns of the OA to obtain the degree of confounding and envisage how main effects and/or two-factor interactions are confounded with other two-factor interactions.

Some control factors (in the control array) are harder to change than others, so these are assigned to columns were changes occur least often. For example, mould temperature and barrel temperature in an injection-moulding process need more time to become thermally stabilized than other factors and therefore these factors should be assigned to columns where changes occur least often.

Step 7   Experimental preparation and run.   It was recognized that the experimental preparation involves those activities that occur prior to actual running of the experiment. Thorough preparation is critical to the success of any industrial designed experiment. Poor preparation is the most frequent cause of inconclusive results. Errors in the experimental procedure at this stage can affect the experimental validity. Several considerations were recognized as being recommended prior to carrying out an experiment, such as:

- Selection of suitable location for carrying out the experiment: for any industrial experiment, it is vital to select an appropriate location which is unaffected by external sources of noise. These external sources of noise (e.g. humidity, vibration, type of material, dust, material composition, etc.) would severely affect the experimental results. The experimental environment should ideally be as close to an exact replicate of the user's environment.
- Availability of experimental resources: the necessary equipment, machines, operators, etc. required for carrying out the experiment should be in place.

- Cost–benefit analysis: this analysis is used for assessing the viability of an action in monetary terms. Here the costs of taking a particular action are compared to the benefits achievable from the outcome of an event. In industrial experiments, the total cost of the experiment (which includes material cost, machine time, labour cost, data analysis) can be excessive. An assessment must be made as to the value of the experiment: are the benefits expected to be derived from the experiment worth the money that will be spent?
- Preparation of experimental data sheets: experimental data sheets must be constructed to minimize the opportunity for errors in setting the proper levels for various factors in each trial. The data sheet (i.e. coded and uncoded design matrices and response tables) typically should list the levels for each factor, date and time of the test. It should also have space for the experimenters to record measurements and comments about the test. Recording the time and date can help the team answering questions during the analysis stage such as, "Was there any significant difference in the results of the trials conducted in the morning and in the afternoon, evening or even next day?"
- An overview of the mission statement: a final overview on the purpose and objective of the experiment is very useful so that the whole team knows the logistics of how the experiment will be run, how the data should be recorded, etc. The management should have been informed throughout the planning process of the experiment. The team should also inform the personnel in the area where the experiment is to be run. For example, if the experiment is to be performed in production, the team should inform all production operators who will be involved, seek their concerns, and then modify the experiment accordingly.

The following steps may be considered while performing the experiment in order to ensure that the experiment is performed according to the prepared design matrix (which shows all the combinations of factors to be studied at different levels) and to detect any discrepancies while monitoring the experimental runs.

1. Run the experiment according to the selected factor levels from step 4. The person responsible for the experiment should be present throughout the experiment. It is also recommended to select one operator for the entire experiment to reduce operator-to-operator variability.

2. Ensure that the experimental trials are made according to the chosen design matrix.

3. Monitor the experimental trials: this is to find any discrepancies while running the experiment. It is advisable to stop running the experiment if any discrepancies are found.

4. Randomization of experimental trials: while performing any industrial experiment, randomization is critical to assure that bias is avoided during data gathering. Therefore randomization strategies should be considered (if possible) while running the experiment. Whether to randomize the order in which the matrix settings is run depends on two main considerations: the cost of randomization and whether a time-dependent factor (known or unknown) will disturb the result. For example, if the cost of randomization may be high because it takes a very long time to reach and stabilize a setting, such as temperature, the experiment may be run such that the number of temperature changes is minimized. On the other hand, randomized order may be needed if the environment is be large enough to strongly influence the results. The authors strongly advocate randomization of experimental trials (if possible) as it averages out evenly the effect of all hidden uncontrollable variables. Randomization is quite useful when the brainstorming process fails to identify some factors that cannot be controlled during the experiment.

5. The experimental trials must be replicated (or repeated) if it is possible to do so.

Step 8   Analysing the experimental data.   A short meeting should be held at the conclusion of the experiment with all people involved in the execution of the experiment. General perceptions, unusual occurrences (if any) and discrepancies must be noted and discussed. If the experiment has been designed correctly and performed in accordance with the appropriate design, then statistical analysis will provide the experimenter with statistically valid conclusions.

The methodology provides a systematic and structured procedure for analysing the data obtained from Taguchi experiments. The procedure is presented in Chapter 9. Analysis methods include the analysis of variance (ANOVA) for identifying the main and interaction effects, the signal-to-noise ratio (SNR) for achieving process robustness and, finally, the prediction of response at the optimum condition.

The methodology also provides modern graphical aids such as main effect plots, interaction graphs, SNR plots, normal probability plots and so on. These graphical tools will yield a better understanding of the results to non-statisticians or engineers with limited statistical skills.

Step 9   Interpretation and experimental conclusions.   The purpose of interpretation for designed experiments is to assist engineers in understanding the results obtained from statistical analysis and to take appropriate actions. It is carried out once the data has been analysed using statistical analysis methods. The methodology accentuates the role of interpretation by using sets of rules based on the work of statistical experts and encouraging the use of graphical tools for rapid and easier understanding of results.

If the objective of the experiment defined by the user has been achieved at this stage, the methodology then suggests that the user should go to step 10. On the other hand, if the objective of the experiment has not been met, it is suggested that the user should go to step 3, step 4, step 5 or step 6. For example, if the ANOVA shows that none of the factors or process parameters chosen for a certain process are significant, then the user may be guided to go to step 3 and select new factors for the experiment; step 4 and select higher levels for the factors to be studied in the experiment or change the range of factor settings to observe an effect of factors on the quality characteristic (or response) of interest; or step 5 to re-select the response of interest due to the inappropriate choice of quality characteristic for the first experiment. On the other hand, if three out of seven factors are significant and if the user wishes to perform a full factorial experiment with three factors, then the user is guided to go to step 6 in the methodology. The procedure is repeated as above.

Having interpreted the results of the analysis, it is advisable to ensure that the experimental conclusions are supported by the data, and that they are meaningful in the stakeholder's world. The experimental conclusions will generally fall into one of the following three categories:

- clear, unambiguous supporting conclusions (the objective is achievable in this case and the goal is supportable);
- strong denial of hypothesis (the objective is not achievable, experimental goals need to be re-adjusted and an alternative objective needs to be prepared);
- weak conclusions (cannot support or deny the hypothesis and the objective of the experiment remains questionable). Weak conclusions can be avoided by practising good experimental planning.

Step 10 Confirmatory run/experiment (or follow-up experiment). A confirmatory run/experiment (or follow-up experiment) is necessary in order to verify the results from the statistical analysis. A confirmatory run/experiment should be carried out to confirm the optimal factor settings obtained from step 9. This is to demonstrate that the factors and levels chosen for the influential factors do provide the desired results. The insignificant factors should be set at their economic level during the confirmation run/experiment. If conclusive results have been obtained, improvement action on the product or process under investigation is recommended. On the other hand, if the results do not turn out as expected, further investigation may be required.

In industrial experiments, once the solution has been implemented, it is recommended to monitor the process by constructing control charts on the experiment's response variable(s) and critical factors that influence the response. Control charting will ensure that the problem does not recur.

6.3 Comments

This chapter presents a generic methodology for Taguchi design of experiments as a powerful tool for continuous quality improvement. It discusses the 10-step approach in detail and accentuates the need for a dynamic approach to the design of experiments. The methodology is a guide to engineers with limited statistical knowledge to use Taguchi methods for solving quality problems in manufacturing. The biggest advantage of the methodology is that it is accessible to engineers in a methodical manner and therefore can be implemented easily.

Exercises

6.1 Discuss the problems in the existing methodology for Taguchi design of experiments.

6.2 During the brainstorming session for a Taguchi experiment, a large number of factors were identified initially. Discuss the type of information you would need to determine the number of factors for the experiment, and state how you would proceed to select these factors.

6.3 Discuss the main steps in performing an industrial designed experiment.

6.4   What are the different types of quality characteristics for manufacturing process optimization problems? What is the evaluation criterion for each of these characteristics?

6.5   Explain the important factors to be considered prior to carrying out an industrial designed experiment.

6.6   When do we need to randomize experimental trials?

6.7   Why is a confirmatory run/experiment important for process optimization problems?

6.8   Suppose you perform an industrial designed experiment for optimizing a certain manufacturing process. Assume you have identified five parameters for the experiment. Analysis of the experimental data has shown that none of the factors are statistically significant. What conclusions do you reach in this situation?

References

1. Knowlton, J. and Keppinger R. (1993) "The Experimentation Process", Quality Progress, February, pp. 43–7.
2. "A Guide to FMEA and FMECA", BS 5750, Part 5, 1991.
3. Dale, B.G. and Cooper C.D. (1992) "Total Quality and Human Resources", Blackwell Publishers, Oxford.
4. Roy R.K. (1990) "A Primer on the Taguchi Method", Van Nostrand Reinhold.
5. Barker T.B. (1985) "Quality by Experimental Design", Marcel Dekker, Inc.
6. Logothetis N. (1992) "Managing for Total Quality—From Deming to Taguchi and SPC", Prentice Hall.
7. Antony J. et al. (1996) "Sorting out Problems", Manufacturing Engineer, IEE, October, pp. 221–223.
8. Meisel, R.M. (1991) "A Planning Guide for More Successful Experiments", ASQC Quality Congress Transactions, pp. 174–179.
9. Lindeke, R.R. and Liou, Y.A. (1989) "Methods for Optimisation in the Manufacturing system—The Taguchi Method", Journal of Mechanical Working Technology, Vol. 20, pp. 205–218.
10. Leon, R.V., Shoemaker, A., and Tsui, K.-L. (1993) "Discussion on Planning for a Designed Industrial Experiment", Technometrics, Vol. 35, No. 1, pp. 21–24.
11. Barrentine L.B. (1991) "Concepts for R & R Studies", ASQC Quality Press.

7 PROBLEM CLASSIFICATION

7.1 Introduction

The current approach to solve and tackle manufacturing process quality problems in manufacturing companies using design of experiments (DOE) or Taguchi methods (TMs) often starts with a brief statement of the problem. Having defined the problem, the next step is to perform brainstorming along with cause and effect analysis (CEA) to investigate the causes of the problem and identify the factors that influence the problem. A difficulty of this approach, encountered by many experimenters, has been in understanding the nature of the problem, then converting this engineering problem into a statistical problem to enable appropriate design and analysis to be performed.

DOE and TMs are advanced statistical quality improvement techniques for solving manufacturing process or product problems in a wide range of organizations, especially manufacturing. Frequently industrial engineers have failed to use these techniques effectively for solving manufacturing process quality problems due to a lack of understanding of the knowledge and approach required to apply these advanced statistical methods successfully.

This chapter describes an investigation of a pragmatic and practical approach to the above problem by using both statistical and engineering knowledge. This approach classifies manufacturing process problems based on knowledge extracted from many experts, such as industrial consultants in the field of study, and using several case studies. The result is a problem classification framework (PCF) that will assist industrial engineers to define engineering problems in terms suitable for statistical analyses [1]. The aim of such a framework is to help engineers with limited knowledge of DOE and TMs to match their known (or even unknown) problems with corresponding problems in the PCF.

7.2 Tools for the development of the problem classification framework

In order to develop the PCF, the first task was to investigate the tools required. It was also important to understand the applications of these tools in the area of advanced statistical techniques such as DOE and TMs. The development of the problem classification framework was carried out in four stages:

- understanding and analysing the process;
- identification and investigation of the problem;
- prioritization of the problem causes;
- problem classification.

The approach is illustrated in Figure 7.1.

7.2.1 Understanding and analysing the process

Understanding the process is the most important and essential factor for solving process quality problems. For the devised methodology, flow charts are recommended for understanding the process to be investigated and for detecting areas where process quality problems are most likely to occur. These charts can help to define a process so that everyone will have a shared knowledge of the process. The flow chart should be scrutinized as many times as necessary in order to answer questions such as:

- Do any activities require inputs not already shown?
- Do all activities map properly to process inputs and outputs?

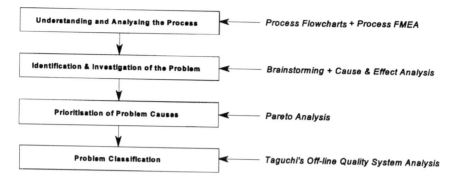

Figure 7.1.   Problem Classification Strategy

- Does the flow chart show all potential paths that the process can take?
- Does the chart capture accurately what really happens?

Having understood the process under study, the next step is to analyse every conceivable way in which a process can cause product failure, and then evaluate the effect of such a failure on the overall process (or system) performance. Process failure mode and effect analysis (PFMEA) is useful to achieve this objective. A PFMEA considers each of the processes involved in the manufacture of the item concerned, what could go wrong, what safeguards exist against the failure, how often it might occur and how it might be eliminated by redesign of the process. PFMEA examines how defects and defectives can arise and reach customers. It does not examine how the product may fail in service due to wear and maloperation. Sound engineering judgement is important in the use of PFMEA for determining the potential process problems and their impact on the overall process performance [2].

For the devised methodology, it was decided that PFMEA should be recommended to determine the severity of the problem under investigation and also to determine the effect of this problem on the output performance of the process.

7.2.2   Identification and investigation of the problem

Problem identification is very critical for any manufacturing process optimization problem, as both the experimental design and analysis aspects are

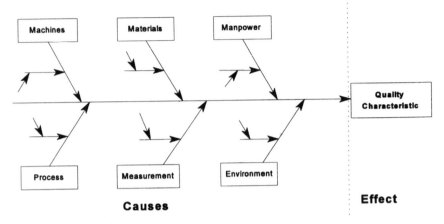

Figure 7.2. Model Cause and Effect Diagram

based on this. For the devised methodology, brainstorming was shown from experts and case studies to be useful for identifying the problem and for identifying the critical factors influencing the problem.

The ideas developed from brainstorming can be applied in conjunction with another problem-solving tool, cause and effect analysis (CEA). This analysis is useful for determining the relationship between the effect of a problem and the cause(s) influencing it. This problem-solving tool is used widely as part of the brainstorming exercise to generate ideas and opinions in order to determine the possible *major causes* of the problem.

Experience has shown that the major causes of any problem in a manufacturing enterprise can be classified into six classes: process, machines, materials, manpower, measurement and environment. Each major cause can be further subdivided into numerous minor causes. For example, a team of people in a certain organization may wish to identify the inputs that are most likely to affect the gas mileage of a vehicle. The minor causes (or sub-causes) under the major cause, say measurement, include: gauge for tire pressure, estimate of full tank, type of speedometer and so on. Figure 7.2 illustrates a typical cause and effect diagram for analysing manufacturing processes.

7.2.3 Prioritization of problem causes

Once the factors influencing the problem are identified, the next step is to prioritize them in order to determine which causes contribute most to a particular problem. Pareto analysis can be used to meet this objective. Pareto analysis is used to separate out the main causes of the problems (or factors) obtained from brainstorming or cause and effect analysis. This generally requires sound engineering knowledge of the process. For any industrial designed experiment, the identification of potential causes or factors influencing the problem play a vital role in its success.

7.2.4 Problem classification

Having prioritized the problem causes (or factors influencing the problem), the next step is to convert a manufacturing process problem into statistical terms from which suitable analysis of the problem takes place. Using the results obtained from a thorough investigation of several industrial case studies, the best approach for achieving this task was to use Taguchi's off-line quality engineering system analysis. The idea behind this approach is to classify a problem defined by the user into one of the three stages of off-line analysis, namely system design, parameter design or/and tolerance design. Taguchi's three-stage approach is used as a basis for the further development of a framework for problem classification.

7.3 Problem classification framework

This section describes how the framework assists industrial engineers in manufacturing companies to classify manufacturing problems. Figure 7.3 illustrates the problem classification framework (PCF). The framework provides a yes/no question to the user at the beginning: 'Has the problem already been identified?' If the answer is no, then the user should study and build a thorough understanding of the process by constructing a flow chart. Once the process has been understood, the user is advised to perform a PFMEA to list problem types based on their severity. Brainstorming can then be used to identify the most critical problem for investigation.

If the answer is yes to 'Has the problem already been identified?', then the user should skip the problem understanding step and the problem identification by brainstorming.

Having identified the problem, the next step is to determine all the possible causes associated with its occurrence. If the causes are not known, they should be identified using CEA. To determine the causes of a specific problem or identify the factors influencing it requires a sound engineering knowledge of the process.

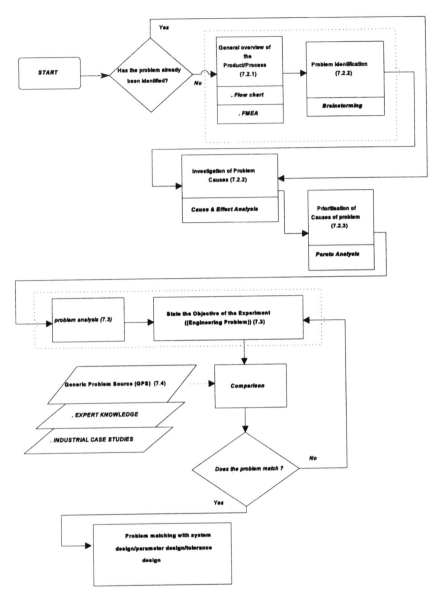

Figure 7.3.   Problem Classification Framework

The next step is to separate out the most significant factors from the less important factors, using Pareto analysis. The user is then in a position to analyse the problem. Problem analysis is used to define the objective of an experiment that will allow the defined problem to be matched against the generic problem source (GPS), where generic problems are contained.

The GPS has been developed using both statistical and engineering knowledge from various experts in the field and from selected case studies of problems in industry. Moreover, the author's expertise and experience in the field of study have also been taken into consideration for the development of the GPS.

The GPS is necessary to provide a means of classifying manufacturing problems in terms of process effectiveness. For example, if the problem is scrap rate at the end of the production line in a certain manufacturing company, then this is obviously an engineering problem. The cause of this problem may be either low-grade materials going into the process, or excessive variability in the process itself. The engineering problem is then converted into a statistical problem, using Taguchi's quality engineering system model, before the appropriate choice of experimental design is made, based on the problem under investigation.

The objective of the experiment defined by the user must be very clear, simple, specific and concise. Having defined the objective of the experiment, the user will then try to match the problem against the generic problems in the GPS. The problem defined by the user can be redefined until problems match each other. When the problem defined by the user matches the problems available in the GPS, then the user will be guided to select an appropriate design from Taguchi's three-stage approach to quality engineering system, from which an experimental analysis of the problem becomes possible.

Table 7.1 shows some of the manufacturing process problems extracted from various case studies to support the development of the GPS. Generic problems, such as minimizing or maximizing response, reducing response variability (increasing precision) or bringing the response mean on target (increasing accuracy) were derived from case studies.

7.4 Generic problem source (GPS)

The GPS encompasses the generic manufacturing process problems and is considered as the heart of the problem classification strategy. Table 7.2 lists the generic problems to match the specific ones given in Table 7.1.

For example, suppose the problem defined by the user is to maximize

Table 7.1.   Manufacturing Process Problems from Industrial Case Studies

Problem number	Nature of the Problem/objective of the experiment
1	To determine the optimal settings for the controllable factors that would produce good adhesion during the assembly of good gaskets to the automobile body
2	To determine the optimal settings of significant process parameters that could minimise the effect of sheet-moulded compound formulation
3	To reduce porosity of a certain reinforced reaction injection moulding process
4	To improve the life of a core tube when subjected to hydraulic test
5	Excessive variation in the surface roughness of an engine part
6	To reduce the force required to assemble the elastomeric connector to the tubing of a certain mechanical product
7	To reduce the post-extrusion shrinkage of a speedometer cable casing on automobiles
8	To maximise the adhesive strength of surface mountain devices to printed circuit boards for a certain SMT
9	To study the effect of critical factors responsible for excessive variation in biscuit length and weight
10	To obtain surface roughness on a certain material using a milling process
11	To minimise shrinkage of plastic parts and to reduce variability in shrinkage
12	Excessive variability in tension of the coil springs used in rocking chairs
13	Excessive variability in automotive shock absorber damping force
14	To identify the factors affecting sensitivity to contamination in an axial piston hydraulic unit
15	To develop a robust screen printing process to achieve print uniformity
16	Higher solder defects rate on circuit board assemblies
17	Excessive variation in closing effort of an automobile glove box assembly
18	To determine the bonder machine settings to maximise the average bond pull strength or shear strength
19	To achieve robust reliability for light emitting diodes
20	To identify the critical substrate process variables affecting wirebonding

Table 7.2.   Generic Problem Source (GPS)

Problem number	Nature of the problem/objective of the experiment
1	To reduce the response variability
2	Rapid process understanding and quick response to management
3	To reduce product development time at the system design stage
4	To improve the reliability of a device/component
5	To identify the most significant factors from a large number of factors
6	To determine the optimal operational settings of a process a) Minimisation problem b) Maximisation problem
7	To set tolerances on the critical design parameters for achieving desired response variability
8	To study the relationship between a set of independent variables and response
9	To reduce the response variability and bringing the process mean on target
10	To assisting with the development of a new process/technology
11	To identify and analyse factors which affect mean response and response variability separately
12	To establish the balance between costs and losses due to variability

the weld strength and to reduce the variability in weld strength distribution. Here the user has defined two objectives, the first being maximizing the weld strength and the second to reduce the variability in weld strength. In other words, the user has to identify those factors that affect the mean response and those that affect the response variability separately.

The next step is to match this problem against the problems in the GPS. The user may select problem number 9 or 11 in the GPS (Table 7.2). If the user already knows a specified target value to be achieved, then problem number 9 is the best choice. Otherwise, the user is guided to select problem number 11. Having completed the problem-matching process, the next step is to assist the user in accommodating the matched problem under any one of the following four types of categories:

- system design;
- parameter design;

- tolerance design; or
- a combination of parameter and tolerance design.

The selection of design stages (as listed above) depends on the nature of the problem and the objective of the experiment. For some problems, the user may have to select a combination of parameter and tolerance design in order to achieve the desired functional performance of processes which could not be achieved through parameter design. In some other situations, the user has to tighten the tolerance limits on selected critical design parameters or process parameters to achieve the required functionality of products. If the mathematical relationship between the response and the independent variables is unknown, then a parameter design experiment must be performed to identify significant parameters or factors influencing the response. For example, if the user has to achieve the desired variability in the performance characteristic (e.g. strength, shrinkage, efficiency) of a product or process, then the user is guided to select a parameter design (assume that the critical parameters affecting the performance characteristic are unknown). On the other hand, if a parameter design has already been performed by the user, then the user is guided to select the tolerance design to achieve the desired variability.

It is important to note that the framework is developed for industrial engineers in manufacturing organizations with limited statistical skills for choosing the suitable design in Taguchi's model for quality engineering system.

7.5 Problem selection framework (PSF)

A problem selection framework (PSF) was developed to assist the user in accommodating the defined problem under any one of the four types of classification, i.e. system design, parameter design, tolerance design or a combination of parameter and tolerance design. The PSF demonstrates what kind of manufacturing process quality problems fall under the above four types of classification. The PSF is shown in Table 7.3. The framework shows the nature of the problems and will assist people in manufacturing companies with limited knowledge in advanced statistical techniques (i.e. design of experiments and Taguchi methods) to match their problems against the categorized problems in the selection framework.

For tolerance design problems, the common practice is to perform a loss function analysis [3] as there is a trade-off between quality loss and costs associated with the components or parts. The results from various industrial

Table 7.3. Problem Selection Framework × – Not applicable ✓ – Applicable

Problem number	Nature of the problem/objective of the experiment	S.D.	P.D.	T.D.	P.D. + T.D.
1	To identify the most significant factors from a large number of factors	✓	✓	×	×
2	To reduce the response variability	×	✓	×	×
3	To set tolerances on the critical design parameters for achieving desired response variability (design parameters are unknown)	×	×	×	✓
4	To set tolerances on the critical design parameters for achieving desired response variability (design parameters are known)	×	×	✓	×
5	Rapid understanding of the existing process and providing a quick response to management	×	✓	×	×
6	To reduce the response variability and bring the mean response on target value	×	✓	×	×
7	To assist with the development of a new process or technology	✓	×	×	×
8	To study the relationship between a set of independent variables and the response variable	✓	✓	×	×
9	To determine the optimal settings of a process a) Minimisation problem b) Maximisation problem (for the process under development)	✓	×	×	×
10	To determine the optimal settings of a process a) Minimisation problem b) Maximisation problem (for the existing process)	×	✓	×	×
11	To analyse the factors which affect the mean response and response variability separately (for the existing process)	×	✓	×	×

case studies have assisted to improve the PSF; especially in terms of classifying and discriminating problems under parameter and tolerance design. For example, if the user has to achieve the desired variability in the functional performance characteristic (i.e. strength, shrinkage, life, etc.)

of a product or process, then the user is guided to select a parameter design (assume that the critical parameters are unknown to the user). On the other hand, if a parameter design has already been performed by the user, then the user is guided to select the tolerance design to achieve the desired variability.

The following section will take the form of real case examples from two manufacturing companies in order to validate the entire approach explained above.

Case 1 Laser welding process

The following steps are involved in the problem classification stage:

(a) Nature of the problem

Short life of a certain product due to poor welding process quality. As the nature of the problem was already known, the user was encouraged to skip the PFMEA and brainstorming based on the PCF.

(b) Investigation of problem causes

CEA was used to identify the potential causes (or factors) of the problem which might have an impact on the problem under investigation. Figure 7.4 illustrates the CEA of the problem.

(c) Pareto analysis

Pareto analysis was performed to prioritise the causes of problems (or factors). Five factors out of 12 were selected for the study. The factors selected may then be classified into control, noise and signal factors.

(d) Problem analysis

In this step, the user will analyse the nature of the problem under study so that the objective of the experiment could be matched against the generic problem source (GPS) where generic problems are contained. The GPS is necessary to provide a means of classifying manufacturing problems in terms of process effectiveness.

(e) Objective of the experiment

The user has classified the objective of the experiment into two stages:

1. To improve the life of the core tubes, i.e. to identify the factors or process parameters that affect the mean life of the core tubes.
2. To reduce variability in the life of core tubes, i.e. to identify factors that affect the variance or standard deviation of response (i.e. life).

In other words, the user wanted to identify the factors that affect the mean response and response variability.

(f) Problem matching

In this step, the user will try to match the defined objective against the generic problems in the GPS. In this particular case, the problem defined by the user matched with problem number 11 in Table 7.2.

(g) Statistical problem classification

This step was used to convert the above engineering problem into a statistical problem so that an appropriate choice of orthogonal array design and analysis of the problem could take place. As the problem is linked to the existing process, the user is guided to select the parameter design (refer to Table 7.3).

The next step was to determine the influence of each of the five factors and the anticipated interactions among the factors with a certain degree of statistical confidence. As the user wanted to study five main factor effects and four two-factor interactions; an L_{16} OA was selected.

Case 2 Injection moulding process

The following steps are involved in the problem classification stage:

(a) Nature of the problem

Customer dissatisfaction due to part shrinkage that occurs after curing. The problem was already established and therefore the user was encouraged to skip PFMEA and brainstorming.

(b) nvestigation of problem causes

Part shrinkage is the amount of measured deviation from the desired part size. Figure 7.5 illustrates the CEA of the problem. The factors

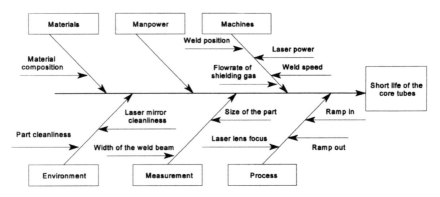

Figure 7.4.   Cause and Effect Analysis for the problem

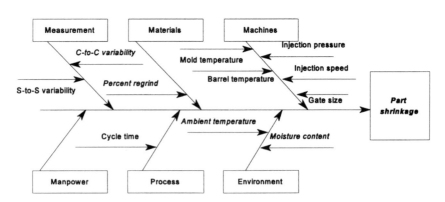

Figure 7.5.   Cause and Effect Analysis for Injection Moulding Process

that are shown in italics are those determined to be uncontrollable in actual production; however, with special care, they were controlled in the experiment.

(c) Pareto analysis

Pareto analysis was used to separate out the main factor effects from the trivial many. For the present example, 7 out of 11 were of interest to the experimenter.

(d) Problem analysis

The problem was analysed to define the objective of the experiment so that the user's defined problem could be matched against the problems in the GPS.

(e) Objective of the experiment

The objective of the experiment was to minimize the part shrinkage problem and therefore to improve customer satisfaction. In other words, the user wanted to identify the factor settings that will optimize the response (i.e. shrinkage). As the quality characteristic (or response) for the experiment was shrinkage, obviously the user wanted to minimize the response.

(f) Problem matching

In this step, the user will try to match the defined objective against the generic problems in the GPS. For the present example, the problem defined by the user matched with problem number 6 (a) in Table 7.2.

(g) Statistical problem classification

As the problem is linked to the existing process, the user is guided to select the parameter design in Taguchi's model of quality engineering system

(Table 7.3). Having selected the parameter design, the next step was to determine the impact of each of the factors chosen for the experiment and then the optimal settings of these factors to minimize the part shrinkage. As the user wanted to study seven main factor effects and three two-factor interactions; an L_{16} OA was selected.

7.6 Conclusions

This chapter illustrates the development of a strategic and practical approach to classifying manufacturing process quality problems using Taguchi's off-line quality engineering system. A problem classification framework (PCF) was developed with the aim of guiding industrial engineers with limited knowledge in advanced statistical techniques to match their defined problems with the problems in the classification framework. This is followed by the selection of an appropriate choice of experimental design from which experimentation and statistical analysis of the problem takes place.

At the outset of the research, three categories of problem classification were defined: system design, parameter design and tolerance design. One of the problems encountered was the classification of problems under a combination of parameter and tolerance design. This was resolved after much expert advice and consultation with practising consultants. The resulting problem selection framework (PSF) was thus developed. Having selected which Taguchi approach (i.e. system design, parameter design, tolerance design or a combination of parameter and tolerance design) is required, the user is then guided to choose the most appropriate orthogonal array design for the problem. This is then followed by the statistical analysis and interpretation of results so that necessary actions can be made for continuous improvement.

Exercises

7.1   What is a problem classification framework?
7.2   Discuss the tools and techniques needed for the problem classification framework.
7.3   Explain the importance of a generic problem source in solving process quality problems.
7.4   What do you mean by a problem selection framework?
7.5   An experimenter is interested in reducing the variability in pull

strength for a certain wire-bonding process. The objective of the experiment was to identify factors that contribute to excessive variability and then determine the optimal factor settings that yield minimum variability in pull strength. Assume that four out of six factors and an interaction effect was of interest to the experimenter. Each factor was studied at two levels. How would you analyse this problem?

References

1. Antony, J. and Kaye, M., "Sorting Out Problems", Manufacturing Engineer Journal, IEE, Vol. 75, No. 5, October 1996, pp. 221–223.
2. Peace, G.S., "Taguchi Methods—A Hands-on Approach", Addison-Wesley Publishing Company, 1993.
3. Antony, J., "Likes and Dislikes of Taguchi Methods", Journal of Productivity, Vol. 37, No. 3, October–December, 1996, pp. 477–481.

8 METROLOGY CONSIDERATIONS FOR INDUSTRIAL EXPERIMENTATION

8.1 Introduction

In the most simplified manner, metrology is the science of measurement. Measurement is an operation of determining the value of a quantity. For any measurement to be of use, it must satisfy the following two criteria:

- the system of measurement must be recognized and used by others in the field of application; and
- the degree of accuracy of the measurement must be known.

For industrial experiments, the response variable or dependent variable will have to be measured either by direct, indirect or comparative methods. These measurement methods, however, produce variation in the output or response. When evaluating process performance, some of the common sources of variation are the measurement process, the measuring equipment and the test method itself. This chapter discusses the method of measurement, types of errors in measurements, the importance of measurement system accuracy and precision and sources of variation in the measurement process.

8.2 Method of measurement

The commonly used methods of measurement are:

- direct method;
- indirect method; and
- comparison method.

8.2.1 Direct method of measurement

This is a method of measurement in which the value of a quantity to be measured is obtained directly [1], e.g. measurement of current using an ammeter, measurement of length using a calliper, etc.

8.2.2 Indirect method of measurement

This is a method of measurement in which the value of a quantity is obtained from measurements made by direct methods of other quantities linked to the quantity to be measured by a known relationship. Measurement of the density of a certain object on the basis of measurements of its mass and its geometrical dimensions is an example of the indirect method of measurement.

8.2.3 Comparison method of measurement

This is a method of measurement based on the comparison of the value of a quantity to be measured with a known value of the same quantity. Measurement of pressure by means of a pressure gauge is an example of this.

8.3 Types of errors in measurements

Any measurement must be made to an acceptable degree of accuracy, but it must be realized that no measurement is exact. There will be a difference between the result of a measurement and the true value of the quantity being measured. There are various types of such differences or errors in measurements, the most common ones being systematic and random errors.

There are no strict definitions of systematic errors, since what is a systematic error for one experiment may not be for another. Very often, they are constant or at least vary slowly over the time required to make a single measurement. A measurement is accurate if it is free from systematic error.

Random errors are those that vary in an unpredictable manner, in magnitude and sign, when a large number of measurements of the same value of a quantity are made under effectively identical conditions. Suppose an engineer uses a vernier calliper to measure the length of a certain specimen, two types of random error might arise during the measurement. One comes from the engineer reading the calliper scale and the other from the calliper itself. It is important to note that no particular random error can be predicted or corrected for. This does not mean that the measuring device could not be improved in such a way as to eliminate some sources of random errors. Once the precision of the measuring device has been estimated, the average magnitude of the errors can be determined.

Errors in measurement are only a small part of the total experimental error. The measuring devices used for measuring the response value(s) in industrial experiments should be checked for accuracy and be traceable to some internationally recognized standard.

8.4 Precision and accuracy

8.4.1 Precision

Precision of a measurement system is the extent to which the system repeats the results when making repeat measurements on the same unit of product [2]. In other words, it is a measure of the scatter of results of several observations and is not related to the true value (Figure 8.1). It is a comparative measure of the observed values and is only a measure of the random errors. **Precision** is expressed quantitatively as the standard deviation of observed values from repeated results under identical conditions. Here true value is an ideal value which could be attained only if all causes of measurement errors were eliminated.

8.4.2 Accuracy

The **accuracy** of a measurement system is defined as the extent to which the average of repeat measurements made on a single unit of product differs

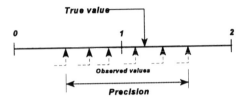

Figure 8.1. Illustration of the definition of precision

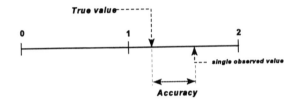

Figure 8.2. Illustration of the definition of accuracy

from the true value. Accuracy is usually expressed in terms of uncertainty. An experiment is said to be accurate or unbiased, if the expected value of the measurement is equal to the true value. Accuracy is a single observed value, and is related to the true value (Figure 8.2).

8.5 Properties of a good measurement system

The measurement system, like all processes, must be stable and capable. A stable and capable measurement is one that is useful for monitoring and improving a process over time. A stable measurement system is often called *reproducible*. A measurement system is said to be *stable* if the measurements do not change over time. In other words, they should not be adversely influenced by operator and environmental changes. A measurement system is *capable* if the measurements are free from bias (i.e. accurate) and sensitive. A capable measurement system requires *sensitivity* (the variation around the average (precision) should be small compared to the specification limits or process spread and accuracy).

Measurements can be either attribute or variable in nature. An **attribute measurement** is one where the results are classified on a pass/fail basis. Examples include separating components as good/bad, checking for missing components and number of components which pass or fail from a certain reliability test. **Variable measurements** are those where the outcomes are reported as a number where numerous values are possible. For example, the tensile strength of a certain specimen in newtons, the belt tension on a certain driving shaft in newton-metres and the viscosity of a certain liquid in poise.

It is important to note that the effect of the average and variation cannot be separated for an attribute output. Moreover, with attribute data, the same level of performance cannot be achieved as with variable data. Variable measurements generally require fewer samples than attributes in order to achieve the same objective.

8.6 The role of measurement in industrial experiments

Industrial engineers and many experimenters in manufacturing organizations are often involved in performing experiments on various core processes. The key to continuous quality improvement is managing processes [3]. The key to managing processes is measurement. Engineers and managers, therefore, must strive to develop useful measurements of their processes. These measurements will be made on the quality characteristic(s) of a product which defines the customer requirements.

The following steps can be of great value when developing a measurement system for industrial experiments.

Step 1   Select the process you want to measure.   This step involves the selection of the process you want to measure and then determination of recipients of the information on measurements, and how it will be used. Important issues that need attention are: "Will the measurement give useful feedback in monitoring a key process? and will the data reveal opportunities for improvement?"

Step 2   Define the target characteristic.   In this step, the target characteristic to be measured should be defined concisely. The definition must be specific enough to suggest procedures for making the measurement. Some target values of measurement are difficult to define specifically. However, a team of people comprising members from quality engineering, production, process engineering and operators can be useful in defining the target characteristics to be measured for experiments. A suitable indicator of the measurement characteristic or some practical way of producing values to reflect the measurement characteristics must also be identified.

Step 3 Perform a quality check. The following questions should be addressed while developing a measurement system for industrial experiments:

- Are we measuring the right quality characteristic(s) which reflect the customer requirements?
- How accurately can we measure the product or process characteristic for the experiment?
- Does variation in the measurement matter? If so, to what extent?
- Is our measurement system stable and capable?

8.7 Gauge repeatability and reproducibility

Repeatability and reproducibility (R&R) studies analyse the variation of measurements of a gauge and the variation of measurements by operators respectively. Here repeatability refers to the variation in measurements obtained when an operator uses the same gauge several times for measuring the identical characteristic on the same part (or sample). Reproducibility, on the other hand, refers to the variation in measurements obtained when several operators use the same gauge for measuring the identical characteristic on the same part (or sample). It is important to note that R&R does not address the total measurement system.

When a manufacturing process is not capable of meeting specification limits, a measurement study is needed to determine whether the improvement efforts should be made on the manufacturing process or on the measurement process. An essential part of the evaluation of a measurement system is the verification that the measurement system is in a state of statistical control. Only then is the measurement system stable and predicable. It is necessary to carry out an R&R study on any operation to which it is planned to apply statistical process control (SPC). If the gauge is not capable, SPC should be deferred until this is achieved [4].

Every observation of a manufacturing process contains both process variation and measurement variation. Measurement variation can be broken into three components, namely:

- gauge variation;
- operator variation; and
- sampling variation.

R&R studies can be used for evaluating operator variation (reproducibility) and gauge variation (repeatability). Customers require both R&R

studies and process capability [5]. Process capability includes both process variation and measurement variation. As a result, R&R studies should be followed by evaluation of variation within the sample and any other sources of variation. A specific example of variation within the sample is apparent in measurements of surface roughness by a profilometer. The test piece itself is sufficiently variable that if the measurement is made at a random position, the variation within the sample will inflate the estimate of repeatability. We will discuss this issue further at a later stage in this chapter. The key point is to ensure that process variability within the sample does not intrude on an R&R study. Determination of an unsatisfactory R&R should always lead to an evaluation of whether variation within the sample is part of the problem.

8.8   Planning gauge R&R studies

A successful gauge R&R study is one that provides good estimates of the variation in the measurement process and identifies the factors that are most influential to that variation. The following features must be considered when planning for R&R studies:

- How and when will the device be calibrated? The gauge should be calibrated before the R&R study begins and not calibrated again until the study is completed.
- The number of operators needed for the study: it is recommended to have three or four operators for the R&R study. Using more than four operators generally makes the study clumsy and unmanageable. The minimum number of operators required for the study is two. If a single operator performs all the measurements using the same gauge, then there is no operator effect.
- The number of samples required for the study: select adequate samples so that the product of the number of samples and the number of operators is greater than or equal to 15. For example, if the number of operators is three for a certain gauge study, then the minimum number of samples required will be five. If there is only operator for the study, then the number of samples required will be at least 15.
- The number of trials: a trial is one measurement on all the samples by each operator. The minimum number of trials required for a gauge R&R study is two. This allows the variability inherent in the device (repeatability) to be separated from the additional variability contributed by the operators (reproducibility).

8.8.1   Procedure for conducting a gauge R&R study

The following steps can be used when conducting any R&R study on gauges.

1. Ensure that the gauge to be used for measurement has been calibrated.
2. Measure all the samples once by the first operator in random order.
3. Measure all the samples once by the second operator in random order. This is repeated until all the operators involved in the study have measured all the samples once.
4. Repeat steps 2 and 3 for the required number of trials.
5. Use software (e.g. Minitab) to obtain the estimate of repeatability and reproducibility. Here we use the procedure based on a spread of 99% (i.e. 5.15σ from normal distribution tables) as used by General Motors.
6. Analyse the results from R&R and take appropriate actions if necessary.

Here repeatability (sometimes called **equipment variation**, EV) is defined as a 5.15σ range (according to General Motors), estimating the spread that covers 99% of the measurement variation due solely to the gauges. Ford uses a 99.7% (6σ range) interval in some of its literature. Reproducibility (sometimes called **appraiser variation**, AV), on the other hand, is defined as a 5.15σ range, estimating the spread that covers 99% of the measurement variation due solely to the operator. The combination of these two sources of variation is called R&R. The estimates of the respective standard deviations are represented by σ_{EV}, σ_{AV} and $\sigma_{R\&R}$. Having obtained the estimates of equipment variation, appraiser variation and R&R, the next step is to translate these values into percentages of the engineering tolerance. Here engineering tolerance is the difference between upper and lower specification limits for a certain characteristic being measured.

8.8.2   Statistical control charts for analysing the measurement process variation

It is essential for the measurement system to be in statistical control. Random selections of 30 parts (gaskets) are chosen from a certain process and each part is measured by one operator using a particular gauge (in random order) two times. Three operators and five parts were involved in this study of measuring gasket thickness using the gauge. The tolerance for the thickness of these gaskets is ±25 units, for a total tolerance spread of 50 units. The data are shown in Table 8.1.

The authors recommend that the data from the above gauge R&R study can be analysed graphically via X-bar and R charts, as shown in Figure 8.3.

The X-bar chart in Figure 8.3 shows many out-of-control points. The X-bar chart in this situation has an interpretation that is somewhat different from the usual interpretation. The X-bar chart shows the discriminating power of the measuring instrument [6]. Most of the averages for the five

Table 8.1. Gasket Thickness Data for Gauge R & R studies

	Operator 1		Operator 2		Operator 3	
Part no.	Trial 1	Trial 2	Trial 1	Trial 2	Trial 1	Trial 2
1	68	62	56	58	53	54
2	108	110	104	100	105	113
3	84	89	85	83	75	81
4	92	94	80	84	84	85
5	63	54	54	65	59	51

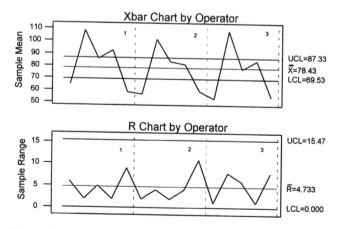

Figure 8.3. Control charts for the Gauge R & R analysis

parts should be outside the control limits of the X-bar chart. If they are not, there would be a serious problem of lack of **measurement discrimination**, i.e. the ability of the gauge or the measuring device to distinguish between units of product.

The range chart (or R chart) shows the magnitude of measurement error, or the gauge capability, often called gauge repeatability. The R chart shows the inconsistency of the variation in the measurement process. If the R chart is out of control, then there is some problem with the measurement process and the problem must be investigated. In the present example, the R chart is in statistical control. This implies that the operators are not having any difficulties in making consistent measurements. The R chart also suggests that there may be a difference between the operators in that operator 3 has larger variation around the mean range ($\overline{R}_1$) than the other two operators (Figure 8.3).

A gauge chart (Figure 8.4) is constructed with the idea of comparing both the variation between measurements made by each operator, and differences in measured values between operators. The chart is usually constructed by plotting the measurement values made by different operators and part numbers on which the measurements were made. A gauge chart is a tool that can be used to assess the differences in measurements between different operators and parts. In Figure 8.4, the reference line is the mean of all observations. The dominant factors here are the part-to-part

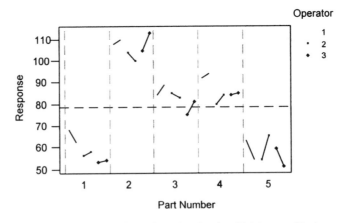

Figure 8.4.   Gauge Run chart for Gasket Thickness Study

variation and the variation in measurements between operators for the same part. Some other interesting patterns also appeared for this study. For example, the second measurement made by operator 3 is consistently higher than the first measurement (excluding the measurement made on part 5).

The X-bar and R-chart methodology breaks down the measurement system variation into repeatability (equipment variation) and reproducibility (operator variation), but does not break reproducibility further into its operator and operator-by-part components of variation. In this case, reproducibility equals just operator variation. The solution to this problem can be resolved by using analysis of variance (ANOVA), which often provides a more accurate assessment of the measurement system than the X-bar and R-chart methods. ANOVA is a powerful statistical technique that analyses the variation of a response and identifies the important contributing factors that cause variation. For gauge R&R study, using ANOVA, the total variation can be divided into four categories: part-to-part variation, operator variation, equipment variation and operator-by-part variation. The operator-by-part interaction cannot be studied using the traditional X-bar and R-chart methodology [7]. Readers who would like to know more about the ANOVA approach on R&R study are recommended to read articles by Montgomery [6] and Wheeler [7].

8.8.3 Analysis of results from R&R studies

The analysis includes the calculation of repeatability, reproducibility and part variation. Having calculated these measures, the next step is to compare these measures with the engineering tolerance. Normally an R&R result of 10% or less would be considered excellent; 11–20% would be considered adequate; 21–30% would be considered marginally acceptable; and greater than 30% would be unacceptable. If the results from R&R studies are not acceptable, it is essential to identify the sources of these results. Is the unacceptability of results primarily due to repeatability (equipment)? Is the unacceptability of results primarily due to reproducibility (operators)? Or is it the combination of both?

If the dominant source of variation is repeatability, one way to improve precision is to average more than one test for the reported measurement. On the other hand, if the dominant source of variation is reproducibility, address training and standardization of procedures. If the dominant source of variation is part variation, the procedure of measuring the parts by different operators needs to be analysed carefully.

Calculations

Repeatability

Average range for the first operator ($\overline{R}_1$) = (6 + 2 + 5 + 2 + 9)/5 = 4.8.

Average range for the second operator ($\overline{R}_2$) = (2 + 4 + 2 + 4 + 11)/5 = 4.6.

Average range for the third operator ($\overline{R}_3$) = (1 + 8 + 6 + 1 + 2)/5 = 3.6.

Repeatability is given by the equation:

$$\text{Repeatability} = \overline{R}_0/d_2$$
$$\text{where,}\ \overline{R}_0 = (\overline{R}_1 + \overline{R}_2 + \overline{R}_3)/3 \tag{8.1}$$

Here d_2 is a constant that can be obtained from any SPC textbooks (refer to Range control chart constants). For repeatability calculations, the value of d_2 depends on the number of repetitions of measurements (or number of trials). In this example, as the number of trials is two, $d_2 = 1.128$. Therefore repeatability = 4.33/1.128 = 3.84 units.

Equipment variation (or gauge repeatability) is defined as 5.15× repeatability = 5.15(3.84). Therefore, equipment variation (EV) = 19.776 units.

Having calculated the equipment variation, the next step is to compare this value against the specified tolerance and express the result as a percentage. The tolerance for the thickness of gaskets is ±25 units. Therefore, we get:

$$\frac{\text{EV}}{\text{Tolerance}} \times 100 = \frac{19.776}{50} \times 100 = 39.55\% \text{ of tolerance.}$$

This means that the repeatability of the measurement device is estimated to be 39.55% of the tolerance (using a spread of 99%). Needless to say, the obtained value of repeatability is considered to be undesirable. The ratio of repeatability (i.e. precision) to that of tolerance is sometimes called the precision-to-tolerance (P/T) ratio.

Reproducibility. For estimating the operator variation (or reproducibility), the first step is to determine the range of the operator averages ($R_{\bar{x}}$). Here, we are dealing with three operators:

average of all measurements for the first operator = $\overline{X}_1$ = 82.4;

average of all measurements for the second operator = $\overline{X}_2$ = 76.9;

average of all measurements for the third operator = $\overline{X}_3$ = 76.0.

$$\text{Range of the operator averages } (R_{\bar{x}}) = \bar{X}_{max} - \bar{X}_{min}$$
$$= 82.4 - 76 = 6.4.$$

$$\text{Reproducibility} = \frac{R_{\bar{x}}}{d_2} \qquad (8.2)$$

Here $d_2 = 1.693$, as the number of operators is equal to three.

Therefore, reproducibility $= \dfrac{6.4}{1.693} = 3.78$ units.

Operator variation is defined as $5.15 \times$ reproducibility $= 5.15(3.78)$.

Therefore, operator variation or appraiser variation (AV) = 19.467 units.

Having calculated the operator variation, the next step is to compare this value against the specified tolerance and express the result as a percentage. For the present example, we get,

$$\frac{AV}{\text{Tolerance}} \times 100 = \frac{19.467}{50} \times 100 = 38.93\% \text{ of tolerance.}$$

This means that the operator effect or reproducibility is estimated to be 38.93% of the tolerance (using a spread of 99%).

Repeatability and reproducibility.   R&R can be combined by squaring, adding and then finding the square root of the final result. For the present example, repeatability = 3.84 units and reproducibility = 3.78 units.

$$\text{Therefore R \& R} = \sqrt{3.78^2 + 3.84^2} = 5.39 \text{ units.}$$

This combined R&R should be multiplied by 5.15, divided by the tolerance and then expressed as a percentage. For the present example, 5.15 R&R = 27.76 units.

When expressed in terms of tolerance, we get, $\dfrac{27.76}{50} \times 100 = 55.52\%$, which means that the combined R&R value consumes about 56% of tolerance.

Part-to-part variation.   The part variation is characterized by finding the range of the part averages, R_p, and then dividing by the appropriate value of d_2. For this example, there are five parts and therefore we will have five average values:

average of measurements for part 1 = $\bar{x}_1 = 58.50$;

average of measurements for part 2 = $\bar{x}_2 = 106.67$;

average of measurements for part 3 = $\bar{x}_3$ = 82.83;

average of measurements for part 4 = $\bar{x}_4$ = 86.50;

average of measurements for part 5 = $\bar{x}_5$ = 59.33;

Range of the part averages = R_p = $\bar{x}_2 - \bar{x}_1$ = 48.17.

$$\text{Part variation} = \frac{R_p}{d_2}$$

where d_2 = 2.326.

$$\text{Part variation} = \frac{48.17}{2.326} = 20.71 \text{ units.} \qquad (8.3)$$

Here the part variation contributes the highest variation and therefore it is important to analyse the method of measurements made on each part by different operators. Moreover, the interaction between operator and part needs to be studied under such circumstances.

8.9 Sampling variation in measurement system analysis

Sampling in the process industries is often done manually and can involve sampling of powdered, liquid or gaseous products [8]. In some measurement situations, variation within the sample cannot be prevented from affecting the R&R study. A good example is the measurement of surface roughness on a certain sample, which may vary significantly across the sample.

The sampling variation can sometimes be quite critical and can be a major component of process capability. If the estimate of repeatability from R&R studies is satisfactory, it means that the variation within the sample is likely to be insignificant. In contrast, if the estimate of repeatability is unsatisfactory or poor, it means that the sampling variation must then be analysed and separated from the estimate of repeatability (or equipment variation). If the sampling variation is significant, then the estimate of repeatability may be in the form:

$$\sigma_{EV} = \sqrt{\sigma_e^2 + \sigma_v^2} \qquad (8.4)$$

where σ_e and σ_v are the estimates of true repeatability and true sampling variance. But it is important to separate the two estimates from equation 8.4. In other words, we need to obtain an independent estimate for σ_e. This can be accomplished by repeating the R&R study, either with special samples where we know there is no variation within the sample, or by specifying exactly where on the production sample, the repeated measurements are to be made. For example, the estimate of equipment variation (or repeatability) from a certain study yields a value of 2.80, an undesirable result. As the sampling variance is mixed or confounded with that of repeatability, we say, $2.80 = \sqrt{\sigma_e^2 + \sigma_v^2}$.

After performing a second study (by controlling the point of measurement), we obtain a new estimate for $\sigma_e = 1.45 = \sigma_{EV}$ (i.e. no sampling variance).

$$\text{Therefore } 2.80 = \sqrt{\sigma_e^2 + \sigma_v^2}$$

$$\text{i.e. } 2.80 = \sqrt{1.45^2 + \sigma_v^2}$$

$$\text{i.e. } 5.7375 = \sigma_v^2$$

$$\text{From this, } \sigma_v = 2.4$$

Here σ_v is an estimate of the variation within the sample, excluding repeatability. If measurements are made at a randomly selected location on the samples, the measurement variation then contains the effect of both repeatability of the gauge used for making measurements and within-sampling variation. On the other hand, if the measurements are made at the same location on each sample, the measurement variation then does not contain the effect of variation within the sample.

8.10   Environmental considerations for measurements

The variation in environmental conditions may contribute to deviations in the readings from actual values. The impact of outer noises (e.g. ambient temperature, relative humidity, supply voltage, etc.) can be applied to measurement devices or gauges which are used for making measurements. These outer noises may cause readings to be sporadically different from actual values. Several considerations can be used to compensate for these noises. For example, if ambient temperature for a certain measurement is a concern and room temperature will remain at a high temperature for a sufficient period of time, then measurements of temperature should be taken at the same time and recorded on the data sheet. Since the noise in this type of situation is close to the same level for each reading, the influence is relatively the same for each reading, therefore deviations from the actual values will be approximately the same for each piece of data.

The ideal situation for dealing with noise is to have measurements taken in a controlled environment. However, it is not feasible and economical to have such controlled environments in many situations. Experienced and skilled operators can assist in developing appropriate techniques for minimizing the impact of these factors during actual production. It is therefore suggested to employ the same operator throughout any measurement process (or even machine settings). Training on the equipment or machines to be used for any measurement should be considered. It is a good idea to be aware of the level of expertise of the operators who actually deal with the equipment.

Exercises

8.1   Discuss the various methods of measurement techniques.
8.2   Explain the role of errors in measurements.
8.3   Exemplify the difference between precision and accuracy.
8.4   Discuss the role of measurements in industrial experiments.
8.5   Differentiate the terms repeatability and reproducibility.
8.6   What is sampling variation in measurement system analysis?

References

1. Keith Brooker (Editor) (1984) "Manual of British Standards in Engineering Metrology", Hutchinson and Co. Ltd.
2. Lewis, E. (1993) "Measurement System Uncertainty", Quality Forum, Vol. 19, No. 1, March, pp. 26–29.
3. Pannell, T.A. and Sundstrom, E. (1990) "Creating the Quality System—the role of measurement in quality system", Tappi Journal, September, pp. 313–314.
4. Bird, D. and Dale, B.G. (1994) "The Misuse and Abuse of SPC: A Case Study Examination", International Journal of Vehicle design, Vol. 15, Nos. 1/2, pp. 99–107.
5. Barrentine, L.B. (1991) "Concepts for R&R Studies", ASQC Quality Press.
6. Montgomery, D.C. and Runger, G.C. (1993–'94) "Quality Engineering", Vol. 6, Part 1, pp. 115–135.
7. Wheeler, D.J. (1992) "Problems with Gauge R&R Studies", ASQC Quality Congress Transactions, pp. 179–185.
8. McNeese, W.H. and Klein, R.A. (1991–'92) "Measurement systems, sampling and process capability", Quality Engineering, Vol. 4, No. 1, pp. 21–39.

9 ANALYSIS AND INTERPRETATION OF DATA FROM TAGUCHI EXPERIMENTS

9.1 Introduction

The statistical analysis and interpretation of data obtained from Taguchi experiments is critical for taking necessary actions and the implementation of these improvement actions for product and process quality improvement. This chapter has been written with the aim of assisting industrial engineers and managers with limited knowledge of mathematics and statistics in analysing and interpreting the data from Taguchi's orthogonal array experiments for process optimization problems. The authors have cut down on the unnecessary equations that tend to confuse managers and engineers with limited mathematical skills. Moreover, they have also tried to include the most essential equations, which are unavoidable, and have presented the results in a graphical manner for easy and rapid understanding. Numerous computer software systems on Taguchi methods are available to enable engineers to analyse their data. The fundamental problem with these software packages is that they provide a black box approach to engineers in statistics. They provide very little support to the interpretation of results obtained from the analysis. As a consequence of this, many industrial engineers would not know what to do next with the results without the help of statisticians. The overall aim of this chapter is to provide adequate guidance and support for engineers with inadequate statistical skills in analysing and interpreting the results from two-level orthogonal array experiments. In order to meet the above objective, the chapter presents a systematic and organized approach for the analysis and interpretation of results.

9.2 Main and interaction effects

The effect (or main effect) of a factor on a response (or output) is defined as the change in response produced by a change in the level of a factor. This can best be illustrated by the following example. In a certain chemical processing plant, an experimenter was interested in studying the effect of two factors, temperature and time, on the process yield. As part of the initial investigation of the process, each factor was kept at two levels. Table 9.1 illustrates the actual factor settings that were used to run the experiment. This is also called an **uncoded design matrix** which displays the levels of each factor for all trials to be run in an experiment. For example, in trial one, temperature and time were set at values 160°C and 15 minutes respectively. For analysis purposes, these actual factor settings should be transformed into coded values, using the following equation:

$$x_i = 2\left[\frac{(f_i - \bar{f_i})}{f_r}\right] \qquad (9.1)$$

where x_i = coded factor setting for factor "i"; f_i = actual setting for factor "i" $\bar{f_i}$ = mean of the actual settings for factor "i" f_r = range of the factor setting.

For example, consider the factor temperature in Table 9.1. The two settings for temperature are 160°C and 180°C. Using equation 9.1, we get,

Table 9.1.   Results from an L_4 OA experiment

Trial number	Time (minutes)	Temperature (°C)	Yield
1	15	160	114
2	25	160	124
3	15	180	108
4	25	180	102

$$x_i = 2\left[\frac{(160-170)}{20}\right] = -1.$$

Similarly,

$$x_i = 2\left[\frac{(180-170)}{20}\right] = +1.$$

Here -1 and $+1$ are the coded low and high levels of temperature. In a similar manner, we obtain -1 and $+1$ coded values for time. On replacing the actual factor settings by coded levels of temperature and time in Table 9.1, we obtain Table 9.2. This is also called a **coded design matrix**.

The coded design matrix shown in Table 9.2 is usually called a 2^2 full factorial design by Western statisticians. In Taguchi's OA design, we generally use 1 and 2 for low and high levels, respectively. Table 9.3 illustrates an L_4 OA design recommended by Taguchi. It is important to note that column 1 in a factorial design (see Table 9.2) is equivalent to column 2 in Taguchi's design (see Table 9.3).

The differences in these experimental designs are simply a matter of

Table 9.2. Coded Design Matrix for the above example

Trial number	Time (minutes)	Temperature (°C)	Yield
1	−1	−1	114
2	+1	−1	124
3	−1	+1	108
4	+1	+1	102

Table 9.3. Taguchi's L_4 Orthogonal Array design

Trial number	Temperature (°C)	Time (minutes)	Yield
1	1	1	102
2	1	2	108
3	2	1	124
4	2	2	114

notation, not substance. This book will follow the notation recommended by Taguchi as many engineers will already have had some introduction to his methods.

Prior to calculating the main effect of a factor, it is important to ensure that the design is balanced and orthogonal. A design is said to be **balanced** when the sum of the coded values for each factor column is equal to zero. Balanced designs are desirable as they simplify the calculations during analysis, and under certain conditions they lend themselves to orthogonal designs. A design is said to be orthogonal if it is balanced and if the sum of the product for all possible variable pairs is zero.

For example, in Table 9.2, the sum of coded values in both time and temperature columns is equal to zero. Moreover, the sum of the product of coded values in both columns is also equal to zero. Therefore, the design matrices shown in Tables 9.2 and 9.3 are balanced and orthogonal. The advantage for orthogonality is that it allows the desired effects to be estimated independently. The effect of a factor can be obtained by the following equation:

$$E_f = \overline{F}_2 - \overline{F}_1 \tag{9.2}$$

where E_f = effect of a factor, $\overline{F}_2$ = average response at level 2, and $\overline{F}_1$ = average response at level 1.

How to estimate the effect of time and temperature on the process yield (refer to Table 9.3):

Effect of time on the process yield.   When time is at level 1 (lower level), the corresponding yield values are 102 and 124.

Therefore, the average yield at level 1 of time, $Ti = \overline{Ti}_1 = (102 + 124)/2$
$$= 113.$$

Similarly, the average yield at level 2 (higher level) of time,
$$Ti = \overline{Ti}_2$$
$$= (108 + 114)/2 = 111.$$

Therefore, effect of time on the yield $= \overline{Ti}_1 - \overline{Ti}_1$
$$= 111 - 113 = -2.$$

Effect of temperature on the process yield.   When temperature is at level 1, the corresponding yield values are 102 and 108.

Average yield at level 1 of temperature, $Te = \overline{Te}_1 = (102 + 108)/2 = 105.$

Similarly, average yield at level 2 of temperature,
$$Te = \overline{Te}_1 = (124 + 114)/2 = 119.$$

Therefore, effect of temperature on the yield $= \overline{Te}_2 - \overline{Te}_1$
$$= 119 - 105 = 14.$$

Interpretation of the main effects.   The relative importance of the two main effects, time and temperature, on the process yield are represented graphically in Figures 9.1 and 9.2 respectively. The sign and magnitude of a main effect tell us the following:

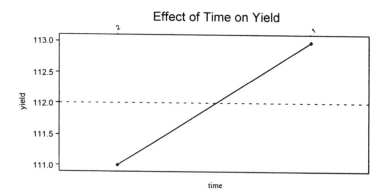

Figure 9.1.   Main Effect of Time

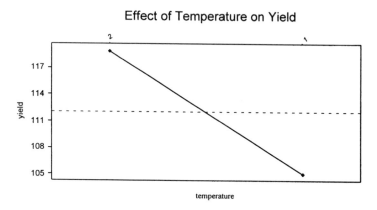

Figure 9.2.   Main Effect Plot of Temperature

- The sign of a main effect tells us of the direction of the effect, i.e. if the response increases or decreases.
- The magnitude tells us of the strength of the effect.

If the effect is positive, it means that the average response at the high level is more than that at the low level. In contrast, if the effect is negative, it means that the average response at the low level is more than at the high level.

As we can see from Figure 9.1, the average yield at low level (level 1) of time is higher than that at high level (level 2). This will give us a negative effect. As the experimenter wanted to improved the process yield, it is clear that level 1 of time is better than level 2. On the other hand, the average yield at the high level (level 2) of temperature is better than that at level 1. Therefore, the factor levels for improving the process yield are: level 2 for temperature and level 1 for time.

Interaction Effect. Two factors are said to interact if the effect of one factor on the response (or output) depends on the level of the other factor. It is the measured change in the response as a result of the combined effect of two or more factors [1]. If the effect of one factor on the response is the same at all levels of the other factor, then the interaction between the factors is zero. For Taguchi-style experiments (or fractional factorial experiments), the interaction effect can be computed using the following equation:

Interaction effect = 1/2 [effect of one factor at high level
of the other factor − effect of the same factor
at low level of the other factor]   (9.3)

For example, P and Q are two factors, each kept at two levels. The interaction between them (represented by either P × Q or PQ) is given by:

P × Q = 1/2 [effect of P at high level of Q − effect of P at low level
of Q] or vice versa.

Here we will illustrate the calculation of the interaction effect between temperature (Te) and time (Ti) on yield (Table 9.3). The interaction plot between temperature and time is shown in Figure 9.3.

$Te \times Ti = 1/2$[effect of Te at high level of Ti - effect of Te at low level of Ti]

$= 1/2[(114 - 108) - (124 - 102)]$ (see Figure 9.3)

$= 1/2[6 - 22]$

$= 1/2[-16] = -8.$

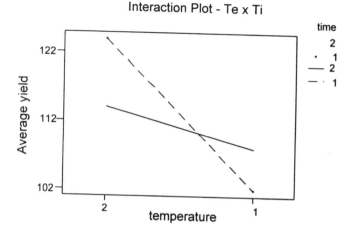

Figure 9.3.   Interaction Graph between Temperature and Time

Table 9.4.   Alternative method to compute the inter-action effect

Temperature (Te)	Time (Ti)	Te × Ti	Yield
1	1	1	102
1	2	2	108
2	1	2	124
2	2	1	114

Alternative methods for calculating the interaction effect.   Consider Table 9.3. In order to compute the interaction effect beween temperature and time ($Te \times Ti$), we need to first multiply columns 2 and 3 in Table 9.3. This is illustrated in Table 9.4.

In Table 9.4, column 3 yields the interaction effect. Having obtained column 3, we then need to calculate the average yield at high (i.e. level 2) and low (i.e. level 1) levels of ($Te \times Ti$). The difference between these will give an estimate of the interaction effect.

$$\therefore \text{Interaction effect } (Te \times Ti) = (Te \times Ti)_2 - (Te \times Ti)_1$$
$$= 1/2 (108 + 124) - 1/2 (102 + 114)$$
$$= 1/2 [232 - 216]$$
$$= 1/2 [16] = 8.$$

For better and rapid understanding of the nature of interactions, the authors have used interaction plots by plotting the response values (or average response values in the case of replications) at each level combination of these factors. An interaction plot displays the average response values at each level combination of factors. If the lines are parallel, then it connotes that there is no interaction between the factors. Non-parallel lines in the interaction plot indicates the existence of an interaction between the factors.

Another way to interpret the interaction effects can be accomplished by constructing a square plot with average response values at each corner of the square. In Figure 9.4 (1,1) implies that both temperature and time are at low levels. (2,1) implies that temperature is at high level and time at low level (i.e. level 1). Therefore, the effect of temperature at the low level of time can be determined as follows:

Effect of temperature at low level of time = 124 − 102 = 22.

Similarly, (1,2) implies that temperature is at low level and time at high level; (2,2) implies that both temperature and time are at low levels. The effect of temperature at high level of time can be determined as follows:

Effect of temperature at high level of time = 114 − 108 = 6.

Here the effect of temperature at two different levels of time (i.e. low and high) is not the same. This connotes that there is an interaction between temperature and time.

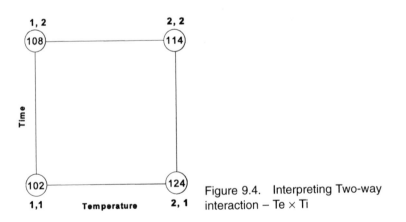

Figure 9.4.   Interpreting Two-way interaction – Te × Ti

Table 9.5. Uncoded design matrix for the welding experiment

Trial no.	Temperature	Flux density	Weld strength		
1	300	10	217	191	252
2	300	20	57	79	77
3	500	10	56	84	76
4	500	20	146	207	234

Table 9.6. Coded design matrix for the welding experiment

Trial no.	Temperature	Flux density	Weld strength		
1	1	1	217	191	252
2	1	2	57	79	77
3	2	1	56	84	76
4	2	2	146	207	234

Scenario

An experimenter wanted to study the effect of welding temperature (°F) and flux density (in./minute) on the resulting weld strength (kg/cm²). As part of an initial investigation, he has studied each factor at two-levels. An L_4 OA was chosen for the study and the results from the experiment are shown in Table 9.5. The purpose of this study is to illustrate how main and interaction effects can be computed for more than one observation or replication. The coded design matrix for the experiment is shown in Table 9.6.

Computation of main effects. For the present study, the main effects are temperature and flux density.

Effect of temperature (T) on weld strength

Average weld strength at low level of temperature

$$= \overline{T}_1 = \frac{(217 + 191 + 252 + 57 + 79 + 77)}{6}$$

$$= 145.5 \, \text{kg/cm}^2.$$

Average weld strength at high level of temperature

$$= \overline{T_2} = \frac{(56 + 84 + 76 + 146 + 207 + 234)}{6}$$

$$= 133.83\, \text{kg/cm}^2.$$

Effect of temperature $(T) = \overline{T_2} - \overline{T_1}$

$$= 133.83 - 145.5$$

$$= -11.67\, \text{kg/cm}^2.$$

Effect of flux density (F) on weld strength

Average weld strength at low level of flux density

$$= \overline{F_1} = \frac{(217 + 191 + 252 + 56 + 84 + 76)}{6}$$

$$= 146\, \text{kg/cm}^2.$$

Average weld strength at high level of flux density

$$= \overline{F_2} = \frac{(57 + 79 + 77 + 146 + 207 + 234)}{6}$$

$$= 133.33\, \text{kg/cm}^2.$$

Effect of flux density $(F) = \overline{F_2} - \overline{F_1}$

$$= 133.33 - 146.0$$

$$= -12.67\, \text{kg/cm}^2.$$

The main effect plots of temperature and flux density are shown in Figure 9.5 and Figure 9.6 respectively.

Computation of interaction effect ($T \times F$). When we multiply columns 2 and 3 in Table 9.6, we get the interaction between temperature, T, and flux density, F as shown below.

T	F	$T \times F$	Weld strength		
1	1	1	217	191	252
1	2	2	57	79	77
2	1	2	56	84	76
2	2	1	146	207	234

Having obtained the interaction column, the next step is to calculate the average response (i.e. weld strength in this example) at low and high levels

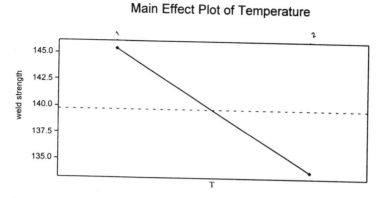

Figure 9.5. Main Effect Plot of Temperature on Weld Strength

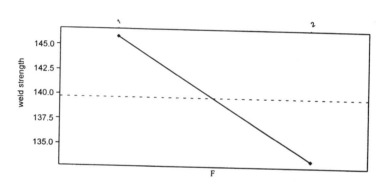

Figure 9.6. Main Effect of Flux Density on Weld Strength

respectively. The difference between the average response values at high and low levels gives the estimate of the interaction effect.

Mean weld strength at low level of $(T \times F) = (T \overline{\times} F)_1$

$= (217 + 191 + 252 + 146 + 207 + 234)/6$

$= 207.83 \, \text{kg/cm}^2.$

Mean weld strength at high level of $(T \times F) = (T \overline{\times} F)_2$

$= (57 + 79 + 77 + 56 + 84 + 76)/6$

$= 71.5 \, \text{kg/cm}^2.$

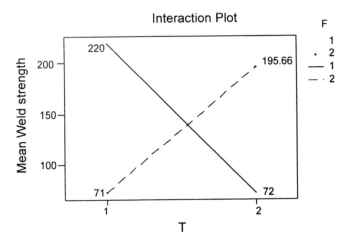

Figure 9.7.   Interaction Plot Between Temperature and
Flux density

Interaction effect $(T \times F) = (T \bar{\times} F)_2 - (T \bar{\times} F)_1$
$$= 71.5 - 207.83$$
$$= -136.33\,\mathrm{kg/cm^2}.$$

The interaction effect between temperature and flux density on the weld
strength is illustrated by an interaction plot (Figure 9.7).

9.3   Determination of the statistical significance of the main and interaction effects

Once the main and interaction effects are estimated, the next step is to
determine which of these effects are statistically significant. Here we
propose both analytical and graphical tools for achieving this objective.
The most powerful analytical tool for identifying the main and interaction
effects is **analysis of variance (ANOVA)**. This is a powerful tool used for
sub-dividing the total variation in the data into useful and meaningful
components of variation. In the context of orthogonal array experiments,
ANOVA is a useful tool to sub-divide the total variation into variation due
to main effects, variation due to interaction effects and variation due to
error. Therefore, mathematically, we can write,

$$\text{Total variation} = V_\mathrm{m} + V_\mathrm{i} + V_\mathrm{e} \qquad (9.4)$$

where V_m = variation due to main effects; V_i = variation due to interaction effects; V_e = error variance

The results of ANOVA are generally shown in a table called an ANOVA table. The ANOVA table generally comprises the following elements:

Source of variation.   This represents the various components of variation that contribute to the total variation. This includes main effects, interaction effects and error. It is important to note that the total variation is the sum of the variation due to the above sources.

Sum of squares.   This represents the sum of squares for each component of variation and the total variation. A sum of squares represents the squared deviation of a random variable from its mean [2]. The sum of squares due to main and interaction effects is a measure of the variability con-tributed by factors or their interactions. For factors at two levels, the sum of squares (SS) due to the main or interaction effects can be computed by the following equation:

$$SS = \frac{N}{4} \cdot [\text{effect}]^2 \qquad (9.5)$$

where, $N = n \times r$ = total number of observations in the experiment; n = number of experimental trials in the orthogonal array; r = number of repetitions or replications; effect = magnitude of main or interaction effect.

The total sum of squares (which is a measure of total variation in the data) can be computed by the following equation:

$$SS_T = \sum_{i=1}^{n} y_i^2 - \frac{T^2}{N} \qquad (9.6)$$

where SS_T = total sum of squares (i.e. total variation); y_i = individual observed response values; T = grand total of the response values; N = total number of response values; $\frac{T^2}{N}$ = correction factor = sum of squares due to the mean.

Degrees of freedom.   The degrees of freedom for each factor is obtained by subtracting one from the number of levels of that factor. The number of degrees of freedom for a factor interaction is obtained by multiplying the degrees of freedom of each factor involved in an interaction effect. The total degrees of freedom is computed by subtracting one from the total number of observed response values for the experiment. Error degrees of freedom is obtained by subtracting the degrees of freedom for main and interaction effects from the total degrees of freedom.

Mean square. The mean square (MS) is obtained by dividing the sum of squares by the number of degrees of freedom associated with the factor effects. Mathematically, we write,

$$MS = SS/v \tag{9.7}$$

where v = degrees of freedom associated with either main or interaction effect.

For factors at two-levels, the mean square due to the factor or interaction effect is the same as the sum of squares, as the number of degrees of freedom associated with the main/interaction effect is unity. The mean square due to error (MSE) is called the error variance.

F-ratio or F-statistic. The F-ratio or F-statistic measures the effect of each factor or interaction relative to the error. In other words, it is the ratio of the mean square due to the factor, or interaction, effect to the error variance (sometimes called the mean square due to error). The usual assumption with regard to the F-ratio is that the individual measurements are independently and normally distributed.

Interpretation of F-ratio. The defining parameters of the F-distribution are the degrees of freedom for each of the quantities in the F-ratio. Thus the F-ratio has two degrees of freedom, denoted by v_1 and v_2 respectively. For OA experiments with factors at two levels, the value of v_1 is always equal to unity whereas v_2 depends on the number of degrees of freedom for error. Given these two parameters and a significance level, α, one can obtain the critical value of F (i.e. F_{critical}) from tables. The critical values of the F-statistic are shown in Appendix C. Here α is the level of significance which gives the probability of the observed factor effect (i.e. main or interaction effect) being due to pure chance. In other words, it is the risk of saying that a factor is significant when in fact it is not. For experimental design, we generally consider two levels of significance; 1% and 5% respectively. If α measures the lack of confidence, then obviously $(1-\alpha)$ measures our confidence for a factor effect to be statistically significant. Therefore 99% and 95% are the most commonly used **confidence levels** in the context of experimental design.

A factor effect is observed to be statistically significant if $F_{\text{calculated}} > F_{\text{critical}}$.

Percent contribution (ρ).   The percent contribution measures the actual percentage variation contributed by a factor or its interaction to the total variation. The percent contribution for a factor or interaction can be obtained by dividing the pure sum of squares (SS') for that factor or interaction by the total sum of squares (SS_T) and multiplying the result by 100. The pure sum of squares (SS') can be computed by the following equation:

$$SS' = SS - [MSE \times v] \qquad (9.8)$$

where SS is the sum of squares due to main or interaction effect, v is the number of degrees of freedom associated with main or interaction effect and MSE is the mean square due to error.

For example, if a factor (say, A) is at two levels, then the product (MSE $\times v$) is the same as MSE. If a factor is at three levels, then we should multiply the MSE by 2, as the degrees of freedom for this factor is two.

The percent contribution (ρ) is then computed by:

$$\rho = \frac{SS'}{SS_T} \times 100. \qquad (9.9)$$

Graphical tools such as Pareto and normal plots are also recommended for the identification of significant main and interaction effects. Moreover, these graphical tools can also be used for supporting ANOVA. We will now illustrate the potential application of ANOVA, main effect and normal plot for studying the effect of four variables on the surface finish of a machined surface.

The four variables were: cutting speed (ft/minute), feed rate (in./rev), nose radius (inches) and side rake angle (degrees). Each factor was kept at two levels. The uncoded design matrix for the experiment is shown in Table 9.7.

Having constructed the uncoded design matrix, we can then construct the coded design matrix (Table 9.8).

Sample calculation of main and interaction effects

Main effect of Nose radius (A)

Mean response at high level of 'A'

$$\overline{A_2} = \frac{(22 + 15 + \text{-----} + 136)}{8} = 67.63$$

Table 9.7. Uncoded Design Matrix for the Surface Finish Experiment

Trial number	Nose radius (A)	Feed rate (B)	Cutting speed (C)	Rake angle (D)	Surface finish (μ inch.)
1	0.015	0.005	600	0	74
2	0.015	0.005	600	10	85
3	0.015	0.005	900	0	56
4	0.015	0.005	900	10	37
5	0.015	0.010	600	0	132
6	0.015	0.010	600	10	284
7	0.015	0.010	900	0	170
8	0.015	0.010	900	10	202
9	0.030	0.005	600	0	22
10	0.030	0.005	600	10	15
11	0.030	0.005	900	0	25
12	0.030	0.005	900	10	13
13	0.030	0.010	600	0	156
14	0.030	0.010	600	10	76
15	0.030	0.010	900	0	98
16	0.030	0.010	900	10	136

Table 9.8. Coded Design Matrix for the Surface Finish Experiment

Trial number	Nose radius (A)	Feed rate (B)	Cutting speed (C)	Rake angle (D)	Surface finish (μ inch.)
1	1	1	1	1	74
2	1	1	1	2	85
3	1	1	2	1	56
4	1	1	2	2	37
5	1	2	1	1	132
6	1	2	1	2	284
7	1	2	2	1	170
8	1	2	2	2	202
9	2	1	1	1	22
10	2	1	1	2	15
11	2	1	2	1	25
12	2	1	2	2	13
13	2	2	1	1	156
14	2	2	1	2	76
15	2	2	2	1	98
16	2	2	2	2	136

Note: "1" represents the low level and "2" represents the high level.

Table 9.9. Estimates of main effects

Main effect	Estimate of effect
A (nose radius)	−62.37
B (feed rate)	115.87
C (cutting speed)	−13.38
D (rake angle)	14.38

Mean response at low level of A $= \overline{A}_1$

$$= \frac{(74 + 85 + \text{-----} + 202)}{8}$$

$$= 130.0$$

$\therefore$ Effect of factor A $= \overline{A}_2 - \overline{A}_1 = 67.63 - 130 = -62.37$

Similarly, we can compute the effects of B, C and D. The estimates of main effects are shown in Table 9.9.

Interaction effect.   Here we illustrate how to compute the interaction effect (A × D) between the factors nose radius (A) and rake angle (D). In order to estimate interaction (A × D), we need to generate a column by multiplying column 2 and column 5 in Table 9.8. In a standard L_{16} OA, factor A must be assigned to column 1 and factor D to column 8. Therefore column 9 yields the interaction between them (refer to standard L_{16} OA in Appendix A).

Mean response at high level of (A × D) $= (A \overline{\times} D)_2$

$$= \frac{(85 + 37 + \text{------} + 156 + 98)}{8}$$

$$= 113.625$$

Mean response at low level of (A × D) $= (A \overline{\times} D)_1$

$$= \frac{(74 + 56 + \text{------} + 76 + 136)}{8}$$

$$= 84.0$$

Interaction effect between A and D $= (A \overline{\times} D)_2 - (A \overline{\times} D)_1$

$$= 113.625 - 84.0$$

$$= 29.625$$

Similarly, we can calculate all the two-factor interactions, such as AB (or A × B), AC, BC, BD and CD. The three-factor and all higher-order interactions can be obtained in a similar manner. Table 9.10 illustrates the estimates of all two-factor interaction effects.

The main effects plot is illustrated in Figure 9.8. The slope of the line implies the magnitude and significance of each main effect. The steeper the slope, the more significant will be the factor effect. As we can see in Figure 9.8, the most dominant factors are nose radius and feed rate. In order to support this claim, we construct the normal probability plot of effects (Figure 9.9).

How to interpret the normal probability plot of effects. The use of normal probability plots is to make judgements about the reality of the observed factor or interaction effects [3]. The factor or interaction effects are plotted along the x-axis and normal scores along the y-axis. The estimated main factor or interaction effects are arranged in order of magnitude (i.e. from smallest to largest). The normal score (or % cumulative probability) can be determined by the formula:

$$P_i = \frac{i - 0.5}{n} \qquad (9.10)$$

where P_i = % cumulative probability corresponding to the ranked factor effects; n = number of effects (main and interaction).

In the normal plot of the effects, points that do not fit the line well usually signal active effects. In other words, active effects tend to appear as extreme points falling off the line. Inactive effects tend to be smaller and fall roughly along a straight line through the origin. In the present example, factors A and B do not fall along the straight line and therefore can be deemed as active effects.

Table 9.10. Estimates of two-factor interaction effects

Interaction effects	Estimate of effects
AB	18.13
AC	−14.13
AD	29.63
BC	−2.88
BD	−21.12
CD	4.63

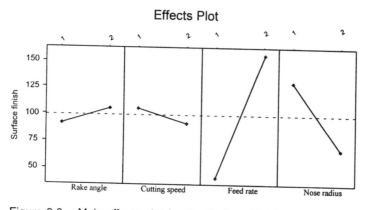

Figure 9.8. Main effects plot for the Surface Finish experiment

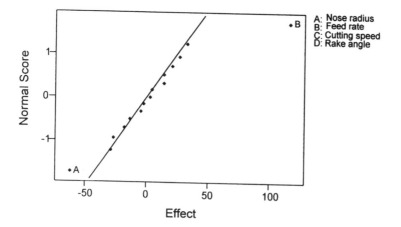

Figure 9.9. Normal Probability plot of main and interaction effects

In the above example, the number of replicates (or repetitions) is zero, i.e. we have one observed value corresponding to each experimental trial. Therefore the total variation is simply the sum of the variance contributed by the main and interaction effects. In such situations, the error degrees of freedom is zero. Therefore the error variance is also equal to zero. This will make the F-ratio undefined. Under these circumstances, we recommend the procedure called **pooling** of insignificant effects. Pooling is a process of obtaining a more accurate estimate of the mean square due to error (or error variance). Pooling is usually accomplished by starting with the smallest sum of squares and continuing with the ones having successively larger effects. This approach is recommended when a factor or interaction effect is determined to be insignificant. Taguchi advocates pooling effects until the error degrees of freedom is approximately equal to half the total degrees of freedom for the experiment [4]. Increasing the degrees of freedom for the error term, as a result of pooling, increases the confidence level of the significant factors. Factor effects (main or interaction) which contribute variation less than 5% of the total variation (i.e. percent contribution in ANOVA) can usually be pooled with the error by adding it to the error sum of squares. The degrees of freedom of these factor effects should also be added to the error degrees of freedom. The authors recommend using a normal plot of effects in order to determine which of the factor effects are insignificant prior to employing pooling.

A larger degrees of freedom naturally results when trial conditions are repeated (or replicated) and standard analysis is performed. When the error degrees of freedom is large, pooling may not be necessary, but to repeat all trial conditions just for the purpose of obtaining a reasonable estimate for the error variance may not be feasible and practical in many situations. However, the authors advocate the repetition or replication of experimental trials, if possible, thus avoiding the process of pooling.

Construction of ANOVA table

Source of variation. The various components of variation are shown in Table 9.11.

Sum of squares. The total sum of squares can be computed using equation 9.6.

$$\therefore SS_T = [74^2 + 85^2 + \ldots\ldots + 136^2] - \frac{1581^2}{16}$$

$$= 88482.44.$$

Sum of squares due to the main and interaction effects can be obtained using equation 9.5. For example, the sum of squares due to nose radius is given by:

$$SS_A = \frac{16}{4} \times [-62.37]^2 = 15560.07.$$

The sum of squares due to interaction AD (or A × D) is given by:

$$SS_{AD} = \frac{16}{4} \times [29.63]^2 = 3511.75.$$

Table 9.11.   ANOVA Table for the Mean Response

Source of variation	Sum of squares	Degrees of freedom	Mean square	F-statistic	Percent contribution
A	15 560.07	1	15 560.07	—	17.59
B	53 703.43	1	53 703.43	—	60.69
C	716.10	1	716.10	—	0.809
D	827.14	1	827.14	—	0.935
AB	1 314.79	1	1 314.79	—	1.490
AC	798.63	1	798.63	—	0.903
AD	3 511.75	1	3 511.75	—	3.969
BC	33.18	1	33.18	—	0.037
BD	1 784.22	1	1 784.22	—	2.016
CD	85.75	1	85.75	—	0.097
All higher order interactions	10 144.69	5	2 028.94	—	11.465
Error	0	0	0	—	—
Total	88 482	15	—	—	—

Because there are no repetitions or replications of experimental point (or trial), we may need to pool the insignificant effects.

The ANOVA table is shown in Table 9.11. The percent contribution is obtained by taking the ratio of the sum of squares due to each main or two-factor interaction to the total variation. The three- and higher-order inter-actions were combined as a single measure and entered into the table. Therefore, the sum of squares due to these higher-order interactions is the sum of four three-factor interactions and one four-factor interaction. Although the percent contribution due to these higher-order interactions is more than 5%, their sum of squares estimate is shared by five sources. Table 9.11 illustrates that the most influential factors affecting the average surface roughness are nose radius and feed rate. Both main effects and normal plots support this argument (Figures 9.8 and 9.9 respectively).

With the error degrees of freedom being equal to zero, the sum of squares due to error and therefore the mean square due to error (or error variance) cannot be determined. Moreover, F-statistics for both the main and interaction effects cannot be calculated since the calculations involve error variance. In order to pursue the calculations, inactive and smaller effects are added together to obtain a non-zero estimate of the error variance. Based on the ANOVA table and normal plots, we now pool or

Table 9.12.   Pooled ANOVA Table

Source of variation	Sum of squares	Degrees of freedom	Mean square	F-statistic	Percent contribution
A	15 560.07	1	15 560.07	10.52	15.92
B	53 703.43	1	53 703.43	36.33	59.02
error	19 218.5	13	1 478.35	—	25.06
Total	88 482	15	—	—	100

From F-tables, $F_{0.01,1,13} = 9.07$, $F_{0.05,1,13} = 4.67$.

combine the insignificant main and interaction effects (i.e. two-factor and higher order) to obtain a new estimate of the error variance, as follows:

pooled degrees of freedom = 13

pooled sum of squares = 19218.5

$$\text{error variance (MSE)} = \frac{19218.5}{13} = 1478.35.$$

The pooled ANOVA table is shown in Table 9.12.

Therefore the critical values of F-statistic at 1% significance level (i.e. 99% confidence level) and 5% significance level (i.e. 95% confidence level) are 9.07 and 4.67, respectively. For factors at two levels, v_1 is always equal to unity and v_2 depends on the number of degrees of freedom for the error term. The calculated F-values from the ANOVA table (see Table 9.12) are greater than the tabled or critical F-values. Therefore, we can conclude that factor effects A (i.e. nose radius) and B (i.e. feed rate) are statistically significant at both 95% and 99% confidence levels (or 5% and 1% significance levels).

9.4   Signal-to-noise ratio (SNR)

Dr Taguchi developed the concept of signal-to-noise ratio (SNR) in quality engineering to evaluate the functional performance of a system. The objective of using the SNR as a performance measure is to develop products and processes insensitive (i.e. robust) to noise factors. The SNR indicates the degree of the predictable performance of a product or process in the presence of noise factors [5]. In other words, the SNR measures the sensitivity of the quality characteristic being investigated in a controlled manner, to those external influencing factors (i.e. noise factors) not under control. The concept of SNR originated in the electrical engineering field. Taguchi had applied this concept effectively to determine the optimal settings for various processes through designed experiments.

The aim of any industrially designed experiment is always to achieve the highest possible SNR, which means that the signal is much higher than the random effects of noise. Process parameter (or factor) settings with highest SNR, always yield the optimum quality with minimum variance. The SNR consolidates several repetitions (i.e. at least two data points are required at each experimental design point or experimental trial) into a single measure that reflects the amount of variation present. Taguchi has developed and defined over 60 different signal-to-noise ratios for engineering applications of parameter design. However, we elucidate only the four most commonly used signal-to-noise ratios for static quality characteristics in industrial processes, these being:

- SNR for smaller-the-better quality characteristics;
- SNR for larger-the-better quality characteristics;
- SNR for nominal-the-best quality characteristics;
- SNR for classified attribute quality characteristics.

SNR for smaller-the-better quality characteristics. For smaller-the-better quality characteristics, the SNR (in decibels) is given by:

$$SNR = -10 \log(MSD) \qquad (9.11)$$

where (mean square deviation), $MSD = \dfrac{\sum y_i^2}{n}$; $i = 1$ to n; y_i = observed response value at each experimental design point (or trial); and n = number of observations in each experimental trial.

SNR for larger-the-better quality characteristics. For larger-the-better quality characteristics, the SNR is given by:

$$SNR = -10 \log(MSD)$$

where $MSD = \dfrac{\sum \left(\dfrac{1}{y_i^2} \right)}{n}$.

SNR for nominal-the-best quality characteristics. For nominal-the-best quality characteristics, two SNRs (SNR_1 and SNR_2) are proposed:

$$SNR_1 = -10\log(s^2) \qquad (9.12)$$

[variance only]. Equation 9.12 is recommended if the objective of the experiment is just to reduce excessive variability in response. Here s^2 is the sample variance at each experimental design point.

$$SNR_2 = 10\log\left(\frac{\bar{v}}{s}\right)^2 \qquad (9.13)$$

[both mean and variance]. Equation 9.13 is recommended if the objective of the experiment is to reduce variability in response and then bring the mean response as close as possible to the target or nominal value. A target value must be specified if this equation is to be used.

For nominal-the-best type 2 quality characteristic, a two-stage optimization strategy is advocated to achieve the objective of the experiment:

- Stage 1   Determine the best settings of the control factors that maximize the SNR (using equation 9.13). In other words, the settings of the control factors that minimize the sensitivity to noise (if any).
- Stage 2   Identify an adjustment factor that has negligible effect on SNR but has a significant effect on the mean response. This adjustment factor can be used to bring the mean response on to the target value. The adjustment factor is sometimes called the signal factor.

SNR for classified attribute quality characteristics. For classified attribute characteristics, the SNR is given by:

$$SNR = -10\log\left[\frac{1}{p} - 1\right] \qquad (9.14)$$

where p is the fraction defective, which can take values between 0 and 1. Frequently, p is expressed as a percentage where it can take values between 0% and 100%. This is also called the omega (Ω) transformation. More information on omega transformation is provided in section 9.12.

9.5 Relationship between the SNR and quality loss function (QLF)

We have defined SNR earlier as:

$$SNR = -10\log(MSD).$$

The relationship between the loss function $\overline{L}(y)$ and MSD is given by:

$$\overline{L}(y) = k.\,MSD \text{ (average loss per unit of the product)}$$

The computations of MSD for different quality characteristics are well explained in Chapter 2. Substituting $MSD = \dfrac{\overline{L}(y)}{k}$ in the SNR equation, we get,

$$SNR = -10\log\frac{\overline{L}(y)}{k}$$

$$\text{i.e.} \ -SNR = 10\log\frac{\overline{L}(y)}{k}$$

$$\frac{-SNR}{10} = \log\frac{\overline{L}(y)}{k}$$

$$10^{\frac{-SNR}{10}} = \frac{\overline{L}(y)}{k} \ (\text{recall power rule: if } \log_b a = x, \text{ then } a = b^x)$$

$$\therefore k \times 10^{\frac{-SNR}{10}} = \overline{L}(y) \tag{9.15}$$

Equation 9.15 yields the relationship between the loss function and SNR.

9.6 When and how to use the SNR analysis

For a single observation for each experimental trial condition (i.e. no repli-cates or repetitions), the standard analysis approach (i.e. analysis of mean response) is recommended. When there are repetitions or replications of the experimental point, whether by outer array designed noise condition or under random noise condition, the SNR analysis must be performed. The SNR has been found to provide a practical way to measure and control the combined influence of the deviation of the population mean from the target and the variation around the mean. The analysis of the Taguichi experi-ments using the SNRs for the observed results can be performed easily using any spreadsheet software.

As SNR is a single measure of variability, the total degrees of freedom of the entire experiment is obtained by: (the number of experimental

trials) −1. The SNR calculation is based on all observations of an experimental trial condition. The SNR values for an experiment are therefore considered as single measures without repetitions. The calculations of main
effects, interaction effects, application of ANOVA, etc. are similar to the
analysis of mean response, as described and illustrated earlier. In other
words, the analysis of SNR is very similar to the analysis of mean response
with no repeated measurements in each trial condition.

It is important to note that the SNR analysis is recommended as long as
the experiment has multiple samples or results for each trial condition, irrespective of whether the experimental layout has a formal outer array or
not. The objective of Taguchi's approach of parameter design is to determine the optimal control factor settings which will dampen the effect of
noise factors (known or unknown) in the process. The method of selecting
the optimal control factor settings which are least sensitive to noise is much
facilitated by the use of SNR analysis.

In order to illustrate how to use the SNR analysis, we will show an
example published in the Taguchi Symposium [6]. An L_8 OA was selected
to study seven factors with the aim of minimizing the out-of-balance condition on a certain wheel cover. This condition may cause vibration of the
vehicle's steering mechanism. For this experiment, it was difficult to identify and include the noise factors for the experiment and therefore repetitions of trials will show the effect of random variation (or noise in Taguchi's
terminology). The experimental layout is shown in Table 9.13. The seven
factors for the study were: mould temperature (A), close time (B), booster
time (C), plunger time (D), back pressure (E), hold pressure (F) and barrel
temperature (G). Five parts per experimental trial were deemed to be a
sufficient number of repetitions needed to estimate the error variance.

As the objective of the experiment was to minimize the out-of-
balance condition, SNR analysis based on smaller-the-better quality char-

Table 9.13. Experimental Layout and Response values

Trial no.	A	B	C	D	E	F	G	Balance (in-oz)				
1	1	1	1	1	1	1	1	0.59	0.47	0.63	0.71	0.59
2	1	1	1	2	2	2	2	0.70	0.91	1.13	0.79	0.78
3	1	2	2	1	1	2	2	0.56	0.44	0.46	0.53	0.46
4	1	2	2	2	2	1	1	1.50	1.55	1.38	1.45	1.45
5	2	1	2	1	2	1	2	1.25	1.36	1.51	1.22	1.25
6	2	1	2	2	1	2	1	1.17	0.97	0.98	0.73	0.97
7	2	2	1	1	2	2	1	1.52	1.40	1.58	1.61	1.57
8	2	2	1	2	1	1	2	0.57	0.51	0.44	0.56	0.44

acteristic was chosen. As we know, the SNR for smaller-the-better quality characteristic is given by:

$$SNR = -10\log(MSD)$$

where $MSD = \dfrac{\sum y_i^2}{n}, i = 1$ to n.

For trial number 1 (see Table 9.13), the MSD can be computed as follows:

$$MSD = [0.59^2 + 0.47^2 + 0.63^2 + 0.71^2 + 0.59^2]/5$$
$$= 0.364.$$

$$SNR = -10\log(0.364) = -10(-0.439) = 4.39.$$

Similarly, the SNR values for the other trials can be obtained. Table 9.14 presents the SNR values for all eight trials. Having obtained the SNR, the next step is to calculate the main effects and interaction effects (if any) based on the SNR values. The calculation of effects for the SNR are analogous to that of mean response.

Sample calculation of effects based on the SNR

For factor A, the average SNR at high level (level 2)

= $(-2.43 + 0.228 + -3.73 + 5.90)/4$

= -0.008.

Similarly, the average SNR at low level (level 1)

= $(4.39 + 1.16 + 6.16 + -3.33)/4$

= 2.095.

$\therefore$ Effect of factor A on the SNR = $-0.008 - 2.095 = -2.103$.

Table 9.14. SNR values for the experiment

Trial number	SNR
1	4.39
2	1.16
3	6.16
4	−3.33
5	−2.431
6	0.228
7	−3.73
8	5.90

Table 9.15.   Effects of the Factors on the SNR

Factors	Average SNR at low level	Average SNR at high level	Effect
A	2.095	−0.008	−2.103
B	0.837	1.25	0.413
C	1.93	0.157	−1.773
D	1.098	0.989	−0.109
E	4.170	−2.083	−6.253
F	1.133	0.954	−0.179
G	−0.611	2.695	3.306

The effects of other factors on the SNR can be calculated in a similar manner. The results are shown in Table 9.15. Having obtained the effects of factors on the SNR, a standard ANOVA can then be performed on the SNR. The aim of such an approach is to identify the factor effects that have a significant effect on the average value of SNR and subsequently reducing variation. The ANOVA approach to the SNR will be discussed in the next section.

9.7   ANOVA for the signal-to-noise ratio

ANOVA for the SNR is very much similar to the analysis of the response with no repetitions or replications at each experimental design point. As the SNR is a single performance measure, pooling of insignificant effects must be performed to obtain adequate degrees of freedom for error. Here we pool the insignificant effects until the degrees of freedom for the error term is nearly half the total degrees of freedom. The ANOVA tables before and after pooling of insignificant effects are shown in Table 9.16 and 9.17 respectively. The tabled (or critical) F-values at 95% (i.e. 5% significance level) and 99% (i.e. 1% significance level) confidence levels are:

$$F_{0.05,1,3} = 10.13, \ F_{0.01,1,3} = 34.12 \ \text{(refer to Appendix C)}.$$

On comparing the tabled and calculated values of F-statistic, we can conclude that factor effects A, C, E and G are statistically significant at both 95% and 99% confidence levels (or 5% and 1% significance levels).

Having performed the ANOVA based on the mean response and the

Table 9.16. ANOVA Table for the SNR (Unpooled)

Source of variation	Degrees of freedom	Sum of squares	Mean square	F-statistic	Percent contribution
A	1	8.85	8.85	—	7.65
B	1	0.341	0.341	—	0.295
C	1	6.28	6.28	—	5.43
D	1	0.024	0.024	—	0.021
E	1	78.20	78.20	—	67.64
F	1	0.064	0.064	—	0.055
G	1	21.86	21.86	—	18.91
error	0	—	—	—	—
Total	7	115.62	—	—	100

Table 9.17. Pooled ANOVA Table for the SNR

Source of variation	Degrees of freedom	Sum of squares	Mean square	F-statistic	Percent contribution
A	1	8.85	8.85	61.88	7.53
B	(1)	0.341	0.341	—	—
C	1	6.28	6.28	43.92	5.31
D	(1)	0.024	0.024	—	—
E	1	78.20	78.20	546.85	67.51
F	(1)	0.064	0.064	—	—
G	1	21.86	21.86	152.86	18.78
pooled error	3	0.429	0.143	—	0.87
Total	7	115.62	—	—	100

Note: Insignificant factor effects are pooled as shown ( ).

SNR, the user (i.e. engineers with limited statistical skills) will then have to answer the question "Are any of the factor or interaction effects significant?" If the answer is no, then it means that:

- the user has chosen the wrong control factors (or process parameters) for the experiment; or
- the user has chosen a wrong response (or quality characteristic) which has no correlation with the selected factors;
- the user has failed to provide adequate control on the effect of noise

(or extraneous factors) which mask the effects of the truly significant effects; or

- the range of control factor settings was too large or too small to observe an effect on the response.

On the other hand, if the answer to the question "Are any of the main or interaction effects significant?" is yes, the user is then advised to determine the optimal process parameter settings.

9.8 Determination of optimal process parameter settings

For Taguchi experiments, it is good practice to perform ANOVA for the mean response and the SNR for identifying the important main and interaction effects. The following three steps were recognized to be useful for determining the optimal process parameter (or control factor) settings.

- **Step 1**   List all the significant main effects and interaction effects (if any) based on ANOVA for the mean response and the SNR. If interaction effects are significant, it is then important to consider all the combinations of factor levels to determine the best combination. For example, for a three-factor (A, B and C) experiment, the ANOVA for the mean response indicates that interaction AC is significant. Assume each factor is kept at two levels: high and low. In this case, it is necessary to look at the following combinations: B_1C_1, B_1C_2, B_2C_1 and B_2C_2. The combination that yields the optimum average response must be chosen for further experiments. The factor levels based on the SNR should also be determined. The factor level combination that yields the maximum SNR is the best combination. If the interaction effects are to be studied, the average SNR at each of the factor level combinations must be determined to arrive at the final optimal factor settings.
- **Step 2**   Check whether or not there is any trade-off in factor levels. If there is, it is recommended that a trade-off analysis using Taguchi's average loss-function must be performed. The average loss-function is calculated for each experimental trial. The factor settings that yield the minimum average loss must be selected.
- **Step 3**   The selection of factor levels for insignificant main effects are based on cost and/or convenience.

Table 9.18. Mean Response Table

Factors	Mean response at level 1	Mean response at high level
A	0.85	1.08
B	0.94	1.00
C	0.88	1.06
D	0.99	0.95
E	0.64	1.30
F	0.97	0.96
G	1.14	0.79

Table 9.19. Pooled ANOVA for the Mean Response

Source of variation	Degrees of freedom	Sum of squares	Mean square	F-statistic	Percent contribution
A	1	0.51	0.51	42.58	7.38
C	1	0.34	0.34	28.25	4.83
E	1	4.31	4.31	357.72	63.27
G	1	1.21	1.21	100.23	17.60
pooled error	35	0.42	0.012	—	6.92
Total	39	6.79	—	—	100

From F-tables, the tabled values of F-ratio are: $F_{0.05,1,35} = 4.13$ and $F_{0.01,1,35} = 7.44$.

We will illustrate here how to select the optimal factor settings for the above experiment (i.e. out-of-balance condition for the wheel cover). Our first task is to construct a mean response table (Table 9.18), which displays the mean response values at low and high level of each factor.

Having constructed the mean response table, the next step is to construct an ANOVA table (Table 9.19) in order to determine which of these effects are statistically significant. On comparing the tabled and calculated values of F-statistic, we may conclude that factors A, C, E and G are statistically significant at both 95% and 99% confidence levels (or 5% and 1% significance levels).

Having identified the significant factor effects, the next step is to determine the optimal settings of these factors which will provide the minimum out-of-balance condition. ANOVA for the mean response and the SNR have shown that factors A, C, E and G are significant. Therefore the first task is to obtain the best factor-level settings based on the mean response. For the present example, the best control factor settings based on the mean response (Table 9.18) are:

- factor A: level 1 (0.85 in-oz);
- factor C: level 1 (0.88 in-oz);
- factor E: level 1 (0.64 in-oz); and
- factor G: level 2 (0.79 in-oz).

This can also be written as $A_1C_1E_1G_2$.

Similarly, we need to determine the best control factor settings based on the average SNR at each level of each factor. The best control factor settings based on the SNR (Table 9.15) are:

- factor A: level 1 (2.095 decibels);
- factor C: level 1 (1.93 decibels);
- factor E: level 1 (4.170 decibels); and
- factor G: level 2 (2.695 decibels).

In other words, we can write: $A_1C_1E_1G_2$.

As there is no trade-off in the factor levels based on the analysis of mean response and also the SNR, the final optimal control factor settings are as follows:

$$A_1B_2C_1D_1E_1F_1G_2.$$

It is important to note that the selection of the levels of insignificant factors were based on **convenience** *and* **economy**. Having determined the optimal factor settings, the next step is to obtain an estimate of the predicted response at the optimal condition.

9.9   Estimation of the response at the optimal condition

The performance at the optimal condition is estimated only from the significant factor or interaction effects. Therefore, the insignificant factor or interaction effects should not be included in the estimate. The selection of factor levels is dependent on the nature of chosen quality characteristic (i.e. smaller-the-better, larger-the-better or nominal-the-best) for the experiment. The procedure depends upon the additivity of the factor effects. If one factor effect can be added to another to predict the response, then

good additivity of factor effects exists. If an interaction effect is involved in the prediction equation, then the additivity between the factor effects is poor.

Consider three factors, say A, B and C, which are important for an experiment. Assume each factor was kept at two levels for the experiment. The analysis has shown that factors A and C were statistically significant. The interaction effects among the factors were identified as insignificant. Now assume that the objective of the experiment was to maximize the response. The optimal factor levels to achieve the objective of the experiment were: $A_2B_1C_2$. Therefore, the estimation of mean response should be based on factors A and C only. The predicted mean response based on the optimal factor levels of A and C can be estimated as follows:

$$\hat{\mu} = \overline{T} + (optimum\ level\ of\ factor\ A - \overline{T}) + (optimum\ level\ of\ factor\ C - \overline{T})$$

$$(9.16)$$

where $\hat{\mu}$ = predicted value of mean response, and $\overline{T}$ = overall mean of all observations in the data.

For the present example, level 2 is the optimal level for both factors A and C. Substituting the mean response values at level 2 of these factors, we get:

$$\hat{\mu} = \overline{T} + (\overline{A}_2 - \overline{T}) + (\overline{C}_2 - \overline{T})$$
$$\hat{\mu} = \overline{A}_2 + \overline{C}_2 - \overline{T}.$$

If we predict the SNR, then we should replace $\hat{\mu}$ by $\hat{SNR}$ In this case, we substitute the mean SNR values at level 2 of each factor. Moreover, $\overline{T}$ is the overall mean of all SNR values.

Now assume that interaction AC was significant for the experiment. If an interaction needs to be incorporated into the prediction equation, then it is important to identify the best combination of AC interaction. Assume A_2C_2 is the best combination. The predicted mean response based on main effects A and C and also interaction AC is given by:

$$\hat{\mu} = \overline{T} + (\overline{A}_2 - \overline{T}) + (\overline{C}_2 - \overline{T}) + [(\overline{A_2C_2} - \overline{T}) - (\overline{A}_2 - \overline{T}) - (\overline{C}_2 - \overline{T})]$$
$$\hat{\mu} = \overline{A_2C_2}.$$

For the wheel-cover experiment, we will illustrate here the computation of the predicted value of mean response (i.e. out-of-balance). The predicted value is given by:

$$\hat{\mu} = \overline{T} + (\overline{A}_1 - \overline{T}) + (\overline{C}_1 - \overline{T}) + (\overline{E}_1 - \overline{T}) + (\overline{G}_2 - \overline{T})$$
$$\hat{\mu} = \overline{A}_1 + \overline{C}_1 + \overline{E}_1 + \overline{G}_2 - 3\overline{T}$$
$$= 0.85 + 0.88 + 0.64 + 0.79 - 3(0.97)$$
$$= 0.25$$

This is the point estimate of the predicted value. Statistically this provides a 50% chance of the true average being greater than $\hat{\mu}$ and a 50% chance of the true average being less than $\hat{\mu}$. The experimenter would prefer to have a range of values within which the true average would be expected to fall with some confidence. Therefore we need to compute a confidence interval for the predicted mean performance at the optimal condition.

9.10   Confidence interval for the estimated value

The confidence level is a maximum and minimum value between which the true average should fall at some stated percentage of confidence. In other words, it is the variation of the estimated result at the optimum condition. Whenever a prediction is made, it is important to establish the confidence interval within which the observed data must lie. That is, we need to conclude positively that our experimental result would be reproducible. A high confidence level may be chosen to reduce risk, but a high confidence level results in a wider confidence interval. The confidence interval for the predicted mean (or estimated) performance at the optimum condition is given by:

$$\text{CI} = \pm \sqrt{\frac{F(\alpha, 1, v_2) \times \text{MSE}}{N_e}} \qquad (9.17)$$

where MSE = error variance; $F(\alpha, 1, v_2)$ = tabled value of F with 1 degree of freedom for the numerator (i.e. for the mean) and v_2 degrees of freedom for the error term; N_e = the effective number of replications.

The effective number of replications is given by:

$$N_e = \frac{N}{1 + v_z} \qquad (9.18)$$

Where N = total number of experiments (or results). This is the case only for the mean response. If we predict the SNR, then N = number of SNR values, as SNR is a single performance measure. v_z = the number of degrees of freedom used in the estimate of the mean (i.e. the degrees of freedom used in predicting the mean performance).

Note that the effective number of replications (N_e) depends on the number of degrees of freedom used to calculate the predicted mean at the optimum condition and does not depend on which factor level is used. Thus, all main and interaction effects used in predicting the response must be included in the degrees of freedom for calculating N_e.

For the wheel-cover experiment, the confidence interval for the predicted mean response (i.e. balance) at the optimum condition can be determined as follows. Assuming a 95% confidence interval (CI) the tabled value of F with 5% significance level, 1 degree of freedom for the numerator and 35 degrees of freedom for the denominator (see Table 9.19) is 4.13. The error mean square or error variance (MSE) is 0.012 (see Table 9.19).

The effective number of replications is given by:

$$N_e = 40/(1 + V_z)$$

v_z = 1 + 1 + 1 + 1 = 4 (as four main effects are involved in the prediction equation).

$$\therefore N_e = 40/5 = 8$$

$$\text{Therefore 95\% CI} = \pm\sqrt{\frac{(4.13)(0.012)}{8}} = \pm0.079$$

Therefore the result at the optimum condition is 0.25 ± 0.079 at the 95% confidence level.

For a 99% confidence level, $F_{0.01,1,35}$ = 7.44; the values of MSE and N_e remain the same.

$$\text{Therefore 99\% CI} = \pm\sqrt{\frac{(7.44)(0.012)}{8}} = \pm0.106$$

Therefore the result at the optimum condition is 0.25 ± 0.106 at the 99% confidence level.

Having determined the confidence interval for the predicted mean response, it is then recommended to conduct a confirmation experiment or run.

9.11   Confirmation run or experiment

Having determined the confidence limits for the predicted mean response, the user is then asked the question, 'Have you achieved the objective of the experiment?' If the answer to the question is yes, a confirmation experiment or run is then performed. The confirmation experiment or run is used to verify that the predicted mean for the factors and levels chosen from the experiment are valid. A selected number of tests are run under the predicted optimum conditions to see whether or not the observed results are closer to the predicted value. If the answer to the question is no, the possible causes for not achieving the objective of the experiment may be:

- wrong design chosen by the user for the experiment;
- incorrect response readings recorded by the user for the analysis;
- lack of expertise of the user in analysing the data;
- lack of expertise of the user in interpreting the data;
- excessive measurement error.

If conclusive results are obtained from the confirmatory experiment, a specific action for improvement must be taken.

Confidence interval for a confirmation experiment. The confidence interval for a confirmation experiment with few samples being taken is given by:

$$CI = \pm\sqrt{F(\alpha,1,v_2) \times MSE \times \left[\frac{1}{N} + \frac{1}{r}\right]} \qquad (9.19)$$

where r = sample size for confirmation experiment, provided $r \neq 0$. As the sample size approaches infinity, then $1/r$ approaches zero and the formula is reduced to that of the confidence interval around the predicted mean response. As r approaches unity, the confidence interval becomes wider.

For the wheel-cover data, a confirmation experiment was conducted. Thirty pieces were produced under the optimal conditions. With $r = 30$, the confidence interval at the 99% confidence level is:

$$CI = \pm\sqrt{(7.44) \times (0.012) \times \left[\frac{1}{8} + \frac{1}{30}\right]} = \pm 0.12.$$

The average response value from the confirmation experiment was estimated to be 0.32, which falls clearly within the 99% CI, showing that the experiment is satisfactory and valid. Moreover, we may infer that the experimental results are reproducible. We have no way of establishing whether additivity is present, prior to performing an experiment. However, if the confidence interval of a confirmation experiment overlaps with the confidence interval of the predicted mean, then we may accept that the results are additive. Here additivity implies that the effect of a factor and the effect of another can be numerically added or subtracted. In contrast, if the confidence interval of a confirmation experiment does not overlap with the confidence interval of the predicted mean, then the result is not additive. In other words, interaction effects are present in the experiment, which implies that the effect of a factor and the effect of another factor cannot be added or subtracted numerically.

9.12 Omega transformation

The prediction equation is based on the assumption that the quality characteristic of interest possesses good additivity. In other words, the magnitude of each of the significant effects can be added together, resulting in an accurate estimate of the total effect. Percentage values such as percent yield, percent loss, percent defective, etc. do not enjoy this characteristic. As the value of the percentage nears 0 or 100, additivity becomes worse and worse. When the predicted value ($\hat{\mu}$) is calculated, a value greater than 100% or less than 0% can be obtained, which is implicitly meaningless. For example, in a certain experiment the objective was to minimize the percentage of bad units. The analysis of the experiment has shown that factors B, D and E are significant. The percentage of bad units at the optimal levels are: $\overline{B}_1 = 8\%$; $\overline{D}_2 = 12\%$; $\overline{E}_1 = 16\%$. The mean percentage of bad units is computed to be approximately 25% (i.e. $\overline{T} = 25\%$). The prediction equation would be as follows:

$$\hat{\mu} = \overline{T} + (\overline{B}_1 - \overline{T}) + (\overline{D}_2 - \overline{T}) + (\overline{E}_1 - \overline{T})$$
$$\hat{\mu} = 25 + (8 - 25) + (12 - 25) + (16 - 25)$$
$$\hat{\mu} = 25 - 17 - 13 - 9 = -14\%.$$

A negative 14% bad has no realistic meaning. Dr Taguchi has advocated the use of the omega transformation (Ω) for converting these percentages into a form that has better additivity. The omega transformation formula is as follows:

$$\Omega = -10 \log \left[\frac{1}{P} - 1 \right] \tag{9.20}$$

where P is the percentage of units within a category of interest. A detailed explanation and derivation of the formula can be obtained in *System of Experimental Design* [4]. The omega transformation converts the percentages between 0 and 1 to values between minus and plus infinity. If we apply the omega transformation to the above data, we will get the following results:

$$\overline{T}_{db} = -10\log\left[\frac{1}{P} - 1\right]$$

$$= \overline{T}_{db} = -10\log\left[\frac{1}{0.25} - 1\right]$$

$$= -4.77\,db\,(\text{decibels}).$$

Similarly, $\overline{B}_{db} = -10.61\,db$; $\overline{D}_{db} = -8.65\,db$; and $\overline{E}_{db} = -7.20\,db$.

Substituting the values into the prediction equation, we get:

$$\hat{\mu} = \overline{T} + (\overline{B}_1 - \overline{T}) + (\overline{D}_2 - \overline{T}) + (\overline{E}_1 - \overline{T})$$
$$\hat{\mu} = (-4.77) + (-10.61 + 4.77) + (-8.65 + 4.77) + (-7.20 + 4.77)$$
$$\hat{\mu} = -16.92\,db.$$

Substituting this value in equation 9.20 and then solving for P, we get,

$$-16.92 = -10\log\left[\frac{1}{P} - 1\right]$$

$$1.692 = \log\left[\frac{1}{P} - 1\right]$$

$$\therefore \left[\frac{1}{P} - 1\right] = 49.20$$

$$P = 1.99\% \approx 2\%.$$

This value is much more realistic and meaningful than the −14% estimated earlier. It is important to note that the omega transformation is most useful when percentage values are very small or very large. The additivity is generally good when the percentages are between 20 and 80%.

9.13 Conclusions

This chapter covered the analysis and interpretation of data from Taguchi orthogonal array experiments. The ultimate purpose of this chapter is to assist engineers with limited statistical skills in analysing and interpreting the results in a organized manner so that necessary actions can then be taken by them. The tools, techniques and equations that are most relevant for the analysis and interpretation were illustrated in a relatively simple manner. Both graphical and analytical techniques were introduced for quick and easy understanding of the results from the analysis.

The steps involved in analysing data from Taguchi-style experiments may be summarized as follows:

- Determination of main effects and interaction effects (if any).
- Construction of main effects and interaction plots.
- Determination of the statistical significance of factor and interaction effects using the analysis of variance (ANOVA), i.e. analysis of mean response. The purpose of this step is to identify those factors that influence the mean response.
- Determination of the optimal settings based on the mean response.
- Identification of the factor and interaction effects (if any) which have an impact on the signal-to-noise ratio (SNR).
- Determination of the optimal settings based on the SNR.
- To check whether or not there is any trade-off in factor levels.
- If there is a trade-off in factor levels, perform a loss function analysis (LFA). Otherwise, determine the optimal factor settings of the process under investigation.
- Perform a confirmation experiment (or run) to verity the optimal settings. If the results are conclusive, take improvement actions on the process. If the results from the confirmation run do not turn out as expected, further investigation may be required.

Exercises

9.1   In a certain casting process for manufacturing jet engine turbine blades, the objective of the experiment is to determine the most significant factor and interaction effects that affect the part shrinkage. The experimenter has selected three factors: mould temperature (A), metal temperature (B) and pour speed (C), each factor being kept at two levels for the study. The response table together with the response values are shown below. Calculate all the main and two-factor interaction effect estimates.

Trial number	A	B	C	Shrinkage values		
1	1	1	1	2.22	2.11	2.14
2	1	1	2	1.42	1.54	1.05
3	1	2	1	2.25	2.31	2.21
4	1	2	2	1.00	1.38	1.19
5	2	1	1	1.73	1.86	1.79
6	2	1	2	2.71	2.45	2.46
7	2	2	1	1.84	1.76	1.70
8	2	2	2	2.27	2.69	2.71

9.2 A welding experiment has been conducted to study the effect of five variables on the heat input in watts [7]. The variables and their settings are shown below.

Variable	Level 1	Level 2
Open-circuit voltage (A)	31	34
Slope (B)	11	6
Electrode melt-off rate (C)	162	137
Electrode diameter (D)	0.045	0.035
Electrode extension (E)	0.375	0.625

The response table for the experiment is shown below. Calculate all the main factor effects. Note that the trials are shown in the randomized order in which they were run.

Trial number	A	B	C	D	E	Response
1	2	1	1	1	1	4141
2	1	2	1	1	1	3790
3	1	2	1	2	2	2319
4	2	2	2	1	1	3765
5	1	2	2	2	1	2488
6	2	2	1	1	2	4061
7	1	2	2	1	2	3507
8	2	1	2	1	2	3425
9	2	2	2	2	2	2450
10	2	1	2	2	1	2466
11	1	1	1	2	1	2580
12	1	1	1	1	2	3318
13	1	1	2	2	2	1925
14	2	1	1	2	2	2450
15	1	1	2	1	1	3431
16	2	2	1	2	1	3067

9.3 An import/export company is working with a construction engineering firm to study the tensile strength of various concrete formulations. Six variables of interest have been identified, which are as follows:

Variable	Level 1	Level 2
Specimen size (A)	2	4
Amount of water (B)	Low	High
Curing time (C)	24	48
Mixing technique (D)	Manual	Machine
Aggregate (E)	Fine	Coarse
Cement concentration (F)	Low	High

An L_8 OA was used to study the effects of six factors on the tensile strength. The results of the experiment are shown in the following table.

Run	C	B	A	D	E	e	F	Response	
1	1	1	1	2	2	2	1	2.3	2.2
2	1	1	2	1	1	2	2	3.5	3.3
3	1	2	1	1	2	1	2	3.0	2.9
4	1	2	2	2	1	1	1	2.1	1.9
5	2	1	1	2	1	1	2	3.5	3.6
6	2	1	2	1	2	1	1	2.6	2.7
7	2	2	1	1	1	2	1	2.9	2.8
8	2	2	2	2	2	2	2	3.9	4.1

Column 7 in the table can be used for estimating pooled error variance.

Construct an ANOVA table to identify which of the variables have a significant effect on the mean tensile strength.

9.4 Suppose you are a process engineer for an aircraft company and are responsible for the painting process. You wish to conduct a designed experiment in order to determine which of the four factors are most responsible in influencing the paint thickness. Moreover, you would like to reduce the variability in paint thickness around the target of 0.5 mm. What should be the optimal factor settings to achieve this objective? The list of factors being tested for the experiment is shown below.

Factor	Level 1	Level 2
Thinner amount (A)	10	20
Fluid pressure (B)	15	30
Temperature (C)	68	78
Vendor (D)	X	Y

It was decided to conduct an eight-run OA experiment with four replicates of each experimental run. The uncoded design matrix and results are shown below.

Trial	A	B	C	D	Paint thickness			
1	10	15	68	X	0.638	0.489	0.541	0.477
2	10	15	78	Y	0.496	0.509	0.513	0.495
3	10	30	68	Y	0.562	0.493	0.529	0.531
4	10	30	78	X	0.564	0.632	0.507	0.757
5	20	15	68	Y	0.659	0.623	0.636	0.584
6	20	15	78	X	0.549	0.457	0.604	0.386
7	20	30	68	X	0.842	0.910	0.657	0.953
8	20	30	78	Y	0.910	0.898	0.913	0.878

9.5   A process engineer is trying to improve the life of a certain cutting tool. He has run an L_8 OA experiment using three factos: cutting speed (A), metal hardness (B) and cutting angle (C). The results obtained from the experiment are shown below. Conduct an ANOVA for the SNR and obtain the combination of factor levels that produces the longest tool life. Also obtain a confidence interval for the predicted mean tool life.

Trial no.	Response	
1	221	311
2	325	435
3	354	348
4	552	472
5	440	453
6	406	377
7	605	500
8	392	419

References

1. Frigon, N. (1994) "Staying on Target with Design of Experiments", Quality Digest, December, pp. 65–69.
2. Mazu, Michael, J. (1990) "Interpreting a Significant Interaction Effectively", ASQC Quality Congress Transactions, pp. 115–120.
3. Benski, H.C. (1989) "Use of a Normality test to Identify Significant Effects in Factorial Designs", Journal of Quality Technology, Vol. 21, No. 3.
4. Taguchi, G. (1987) "System of Experimental Design", Kraus International Publication, UNIPUB—New York.
5. Kapur, K.C. et al. (1988) "Signal-to-Noise ratio Development for Quality Engineering", Quality and Reliability Engineering International, Vol. 4, pp. 133–141.
6. Harper, D. et al. (1987) "Optimisation of Wheel Cover Balance", 5th Taguchi Symposium, ASI, pp. 527–539.
7. Stegner, D.A. et al. (1967) "Prediction of Heat Input for Welding", Welding Journal Research Supplement, Vol. 1, March.

10 INDUSTRIAL CASE STUDIES

10.1 Introduction

This chapter presents a collection of real industrial case studies. These are well planned experiments and not simply a few experimental trials to explore the effects of varying one or more factors at a time. The case studies will also assist both industrial engineers and business managers with limited skills in statistics to conduct industrial experiments in their own organization in order to solve process quality problems. Moreover, this chapter will increase the awareness of the application of experimental design techniques in industry and their potential in tackling process optimization and variability problems.

This chapter presents four industrial case studies with the aim of educating engineers with limited statistical knowledge in selecting the right OA design, followed by analysing and interpreting the data using both graphical and analytical tools and techniques. The results of these case studies have proved to be of great value in refining and improving the devised methodology described in Chapter 6. Each case study will encompass the objective of the experiment (and/or nature of the problem), selection of factors and interactions associated with the experiment, choice of response or quality characteristic, choice of design, analysis and interpretation of data and the results of a confirmation run or experiment.

10.2 Case studies

10.2.1 Optimization of the life of a critical component in a hydraulic valve

Nature of the problem. The life of a critical component in a certain hydraulic valve was short when subjected to a fatigue test. The component assembly is welded and then machined prior to final assembly of the system. The engineering team found that most of the factors affecting the life of the component were related to the welding process. Therefore it was decided to conduct an industrial designed experiment on the welding process with the aim of identifying main and interaction effects and then maximizing the component life and reliability.

Selection of the factors for the experiment. Five control factors and one noise factor at two-levels each were considered for the experiment. The classification of factors into control and noise were achieved using brainstorming. Table 10.1 illustrates the list of control and noise factors considered for the experiment.

Interactions of interest to the experimenter. In order to list all the interactions that were anticipated from the experiment, a sound engineering

Table 10.1. List of factors for the experiment

Name	Factor label	Unit	Factor settings	
			Low	High
Control factors				
Laser power	A	watts	980	1100
Ramp out	B	sec.	2.0	3.0
Ramp in	C	sec.	1.0	2.0
Weld speed	D	rev/sec.	1.0	1.5
Lens focus	E	—	1	2
Noise factor				
Material composition	N	—	N_1	N_2

Note: N_1 = Mn (1.180) and S (0.279), N_2 = Mn (1.159) and S (0.285).

knowledge of the process was highly desirable. For this study, the interactions of interest to the experimenter were: $A \times B$, $C \times D$, $A \times D$ and $B \times D$.

Quality characteristic (or response). The response of interest for the experiment was fatigue life of the component, measured in million cycles.

Type of quality characteristic (or response). Larger-the-better.

Choice of OA design for the experiment. The choice of OA depends on the total degrees of freedom required for the main and interaction effects. The degrees of freedom associated with control factor effects is equal to five and that of the noise factor effect is equal to one. The degrees of freedom required for studying four two-factor interactions is equal to four. Therefore, the total degrees of freedom required for studying the main and interaction effects in the control array is equal to nine. The closest standard OA to achieve this objective is an L_{16} OA, with 15 degrees of freedom. The noise array consists of one noise factor at two levels.

Problem graph and alias structure of the design. The problem graph for the experiment is shown in Figure 10.1. The design generator for the OA design is given by: $E = ABCD$. Therefore the design resolution is five. This means that the main effects are confounded with four-factor interactions or that two-factor interactions are confounded with three-factor interactions. The defining relationship is given by: $I = ABCDE$. The full alias structure of the design is illustrated in Table 10.2.

Response table for the experiment. The response table of the experiment is shown in Table 10.3. The coded design matrix (or standard L_{16} OA) may be obtained from Appendix A. The response table was constructed by assigning factors that were difficult to control during the experiment to the first column, second column and so forth. The response values were obtained based on the experimental conditions at each design point. Each

Table 10.2. Alias structure among the factors

Defining relationship: $I = ABCDE$
Design Resolution: V
Alias structure:

$A = BCDE$	$B = ACDE$	$C = ABDE$	$D = ABCE$	$E = ABCD$
$AB = CDE$	$AC = BDE$	$AD = BCE$	$AE = BCD$	$BC = ADE$
$BD = ACE$	$BE = ACD$	$CD = ABE$	$CE = ABD$	$DE = ABC$

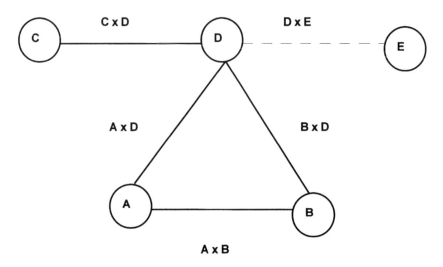

Figure 10.1.   Problem graph for the Experiment

Note: The interactions which are of interest to the experimenter are represented by solid lines and that which is not of interest is represented by a dotted line

combination of factor levels in the control array was run against each noise level in the noise array. Hence the experiment consisted of a total of 32 experimental trials. The experiment was conducted in a randomized manner to distribute evenly the amount of other external extraneous factors which could not be identified during the brainstorming.

Statistical analysis and interpretation.   Having obtained the response values, the first step was to calculate the effect of each factor and interaction. Table 10.4 illustrates the estimated main and interaction effects. As the resolution of the design was V, main effects were clear of two-factor interaction effects and also two-factor interactions were clear of other two-factor interactions.

Once the main and interaction effects are computed, the next step is to conduct an ANOVA on the mean response, in order to identify those main and interaction effects which have a significant effect on the mean response. The pooled ANOVA on the mean response is shown in Table 10.5. In order to support the analytical technique, a half-normal probability plot of factor and interaction effects was constructed (Figure 10.2). The main effects weld speed and ramp out, and two-factor interactions between weld speed and

Table 10.3. Response table for the experiment

Standard order	Trial order	A	B	C	D	E	Response N₁	N₂
							N_1	N_2
1	1	980	2	1	1	1	4.80	1.30
2	15	980	2	1	1.5	2	5.50	6.30
3	9	980	2	2	1	2	5.60	4.80
4	3	980	2	2	1.5	1	9.00	5.60
5	11	980	3	1	1	2	1.60	2.90
6	4	980	3	1	1.5	1	8.40	11.50
7	2	980	3	2	1	1	0.80	4.10
8	13	980	3	2	1.5	2	8.10	8.20
9	10	1100	2	1	1	2	2.00	2.80
10	7	1100	2	1	1.5	1	4.80	5.10
11	8	1100	2	2	1	1	4.70	1.00
12	16	1100	2	2	1.5	2	5.00	3.70
13	5	1100	3	1	1	1	4.60	4.40
14	14	1100	3	1	1.5	2	8.00	8.40
15	12	1100	3	2	1	2	5.00	5.10
16	6	1100	3	2	1.5	1	10.80	8.20

Table 10.4. Table for estimated main and interaction effects

Main and interaction effects	Estimate of effects
Main effects	
A	−0.30
B	1.750
C	0.450
D	3.813
E	−0.388
Interaction effects	
AB	−1.43
AD	0.766
BD	−1.56
CD	−0.370
DE	0.90

Table 10.5.   Table for pooled ANOVA on mean response

Source of variation	Degrees of freedom	Sum of squares	Mean squares	F-statistic	Percent contribution
B	1	24.50	24.5	11.93**	9.67
D	1	116.281	116.281	56.61**	49.22
AB	1	16.245	16.245	7.91**	6.12
BD	1	19.531	19.531	9.51**	7.53
pooled error	27	55.463	2.054	—	27.46
Total	31	232.02	—	—	100

From F-tables: $F_{0.05,1,27} = 4.20$, $F_{0.01,1,27} = 7.66$.
Note: ** implies that both main and interaction effects are statistically significant at 95% and 99% confidence levels (as the calculated F-values are greater than the tabled F-values).

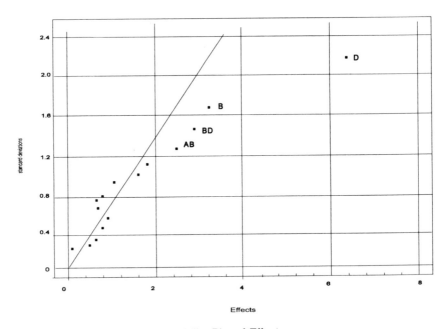

Figure 10.2.   Half-normal Probability Plot of Effects

ramp out, and ramp out and laser power, were judged to be statistically significant. All other main and interaction effects were pooled to obtain a reasonable estimate of error variance.

Having performed the ANOVA on mean response, the next step was to

Table 10.6.

Experimental trial order	SNR
1	4.982
2	15.357
3	14.252
4	16.553
5	5.940
6	19.638
7	0.909
8	18.169
9	7.241
10	13.880
11	2.820
12	12.478
13	13.058
14	18.268
15	14.065
16	19.309

Table 10.7. Main effect and interaction on the SNR

	A	B	C	D	E	AB	AD	ABC	ABD
Level 1	11.974	10.944	12.296	7.907	11.394	14.479	11.251	10.755	13.814
Level 2	12.640	13.670	12.318	16.707	13.219	10.134	13.363	13.856	10.799
Effect	0.666	**2.726**	0.022	**8.780**	1.825	**−4.435**	2.112	**3.101**	**−3.104**

calculate the SNR based on larger-the-better (LTB) quality characteristic. The SNR for LTB quality characteristic was calculated at each experimental design point using equation 9.11:

$$SNR = -10\log(MSD)$$
$$\text{where MSD} = \frac{1}{n} \cdot \sum \left[\frac{1}{y_i^2} \right]; i = 1 \text{ to } n.$$

The SNR values corresponding to each experimental trial condition (Table 10.3) are illustrated in Table 10.6. Having obtained the SNR, the next step was to compute the effect of main and interaction effects on the SNR. These results are illustrated in Table 10.7.

Table 10.7 shows that two main effects, B and D, have significant impact on the SNR. Moreover, one two-factor interaction (AB) and two three-

Table 10.8.   Pooled ANOVA for the SNR

Source of variation	Degrees of freedom	Sum of squares	Mean square	F-statistic	Percent contribution
B	1	29.724	29.724	14.148**	5.025
D	1	308.354	308.354	146.765**	55.713
AB	1	78.677	78.677	37.447**	13.931
AD	1	17.842	17.842	8.492*	2.864
CD	1	15.872	15.872	7.554*	2.505
BD	1	9.647	9.647	4.592	1.373
ABD	1	36.337	36.337	17.295**	6.228
ABC	1	38.539	38.539	18.343**	6.629
pooled error	7	14.705	2.101	—	5.732
Total	15	549.697	—	—	100.00

From F-tables, $F_{0.05,1,7} = 5.59$ and $F_{0.01,1,7} = 12.25$.
** —significant at both 95% and 99% confidence levels (or 5% and 1% significance levels).
* —significant only at 95% confidence level (or 5% significance level).

factor interactions (ABC) and (ABD) also appear to be important. ANOVA was then performed to see which of these effects are statistically significant. The results of the pooled ANOVA are presented in Table 10.8.

Having performed the ANOVA on both the mean response and the SNR, the next step was to identify the optimum condition. The optimum condition is one that provides the optimal factor settings for the process.

Optimal factor settings based on the mean response.   Main effects B and D were significant and interaction effects AB and BD were also significant (based on the ANOVA and half-normal plot, see Table 10.5 and Figure 10.2, respectively). The optimum levels for main effects B and D can be determined as follows: average response at low level (level 1) of factor B = 4.50 (see Table 10.3); average response at high level (level 2) of factor B = 6.25. As we need to maximize the quality characteristic of interest (i.e. life), level 2 is the best. Similarly, for factor D, level 2 is the best.

For interaction effects such as BD, it is important to determine the average response at each of the four combinations: B_1D_1, B_1D_2, B_2D_1 and B_2D_2. Table 10.9 shows the average response values at the above four combinations. The table shows that B_2D_2 is the best combination for maximizing the response function.

Similarly, for interaction AB, the best combination is A_2B_2. The selection of the levels of other factors are dependent on cost and convenience. These

Table 10.9. Average response values

Factor level combinations	*Average response*
B_1D_1	3.375
B_1D_2	5.625
B_2D_1	3.563
B_2D_2	8.950

factors are also called *cost reduction factors*. The optimal factor settings based on the mean response are:

$$A_2B_2C_2D_2E_1$$

Optimal factor settings based on the SNR. The ANOVA for the SNR (Table 10.8) showed that the main effects B and D were significant. Three two-factor interactions and both the three-factor interactions were also found to be significant. The optimum levels for main effects B and D were B_2 and D_2 (see Table 10.7) respectively. For interaction effects, each combination of factor levels must be examined to obtain the best combination. The optimal factor settings based on the SNR (see Table 10.7) are:

$$A_2B_2C_2D_2E_2$$

The two analyses showed that there was no trade-off in factor levels apart from factor E. As factor E is a cost reduction factor, it was decided to set this factor at level 1 (i.e. standard position of lens focus). The final selection of factor settings for the process were:

- laser power: level 2 (1100 W);
- ramp out: level 2 (3.0 s);
- ramp in: level 2 (2.0 s);
- weld speed: level 2 (1.5 rev/s);
- laser lens focus: level 1 (position 1).

Predicted mean life of the component. The predicted mean life of the component based on significant effects was given by:

$$\begin{aligned}
\hat{\mu} &= \overline{T} + (\overline{B_2} - \overline{T}) + (\overline{D_2} - \overline{T}) + [(\overline{A_2B_2} - \overline{T}) - (\overline{A_2} - \overline{T}) - (\overline{B_2} - \overline{T})] \\
&\quad + [(\overline{B_2D_2} - \overline{T}) - (\overline{B_2} - \overline{T}) - (\overline{D_2} - \overline{T})] \\
&= \overline{T} + \overline{A_2B_2} - \overline{A_2} + \overline{B_2D_2} - \overline{B_2} \\
&= 5.375 + 6.875 - 5.225 + 8.95 - 6.26 \\
&= 9.72 \text{ million cycles.}
\end{aligned}$$

95% Confidence interval (CI) for the predicted estimate

$$CI = \pm \sqrt{\frac{F(\alpha, 1, v_2) \times MSE}{N_e}}$$

$F(0.05,1,26) = 4.23$; N_e = effective number of replications = $32/(1 + 4) = 6.4$; MSE = error variance = 2.054 (from Table 10.5).

$$\therefore CI = \pm \sqrt{\frac{(4.23) \times (2.054)}{6.4}}$$
$$= 1.17 \text{ million cycles.}$$

Therefore, the 95% CI for the predicted life = 9.72 ± 1.17 million cycles
$$= (10.89, 8.55).$$

Having determined the optimal factor settings and the predicted mean response, the user was asked to answer the question 'Have you achieved the objective of the experiment?' The mean life of the component at the standard conditions (i.e. process parameter settings based on current production conditions) was 8.00 million cycles and the predicted mean life of the component at the optimal condition was 9.72 million cycles. This showed a significant improvement in the life of the component. In order to verify the results from the prediction equation, confirmation runs were performed at the optimal factor settings.

Confirmation run. Confirmation runs were performed based on the optimal factor settings and showed that the optimum life of the component was equal to 10.10 million cycles. As this value lies in the interval (8.55–10.89), it was concluded that the experiment was sound and satisfactory. The life of the component was therefore improved by over 25% using Taguchi methods of experimental design.

10.2.2 Optimization of welding on cast iron using Taguchi methods

This case study illustrates the role of Taguchi's parameter design for optimizing a process for radiographic quality welding of cast iron.

Objective of the experiment. The objective of the experiment was to determine the optimal control factor settings that gave minimum crack length.

Selection of process output or response. Having identified the objective of the experiment, the next step was to identify a suitable output for the experiment. The response of interest to the experimenter was 'crack length', measured in millimetres.

Selection of the factors and factor interactions. Control factors for the experiment were identified by brainstorming. Employees from production, quality, design, etc. were involved in the brainstorming session. Five control factors were selected to study for the experiment. No noise factors were considered. The team was also interested in studying two interactions. The list of control factors, factor interactions and their levels are shown in Table 10.10.

Type of quality characteristic. Smaller-the-better.

Selection of OA for the experiment. The degrees of freedom required for studying the five main effects is equal to five. The degrees of freedom required for studying the two interaction effects is equal to two. Therefore, the total degrees of freedom required is equal to seven. The closest standard OA for studying seven effects with seven degrees of freedom is an L_8 OA.

Problem graph and alias structure of the design. The problem graph for the experiment is shown in Figure 10.3. As main effects are aliased (or confounded) with two-factor interactions (see Table 10.11), the resolution of the design is III. The design generators, defining relationships and the confounding pattern among the factors are shown in Table 10.11.

Table 10.10. Control factors and interactions of interest

Factors and interactions of interest	Levels	
Main effects	1	2
A — Current (amperes)	110	135
B — Bead length (mm)	20	30
C — Electrode make	1	2
D — V groove angle (degrees)	45	60
E — Welding method	1	2
Two-factor interactions		
Current × Bead length (A × B)	—	
Current × Welding method (A × E)	—	

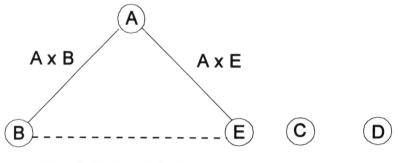

Figure 10.3.   Problem graph for the experiment

Table 10.11.   Confounding relationship among the factors

Design generators: D = AC and E = ABC
Design resolution: III
Defining relationships: I = ACD = ABCE = BDE
Confounding pattern:

A = CD = BCE	B = DE = ACE	C = AD = ABE
D = AC = BE	E = BD = ABC	
AB = CE = ADE	AC = BE	BC = AE = ABD = CDE

Response table for the experiment.   Table 10.12 is the response table for
the experiment. It was constructed by assigning factors that were difficult
to control during the experiment to the first column, second column and
so forth. The response values were recorded corresponding to each
design point.

Statistical analysis and interpretation of results.   Having obtained the
response table, the first step was to calculate the average crack length at
levels 1 and 2 of each factor and interaction effects. The difference will yield
the effect of each factor or interaction. Table 10.13 presents the results.
Figure 10.4 illustrates the main effects plot of factors.

Figure 10.4 shows that factor E has significant impact on the creak length.
In order to support this claim, an ANOVA table for the mean response
(Table 10.14) was constructed. The interaction between current and bead
length also appeared to be significant. The interaction plot between current
and bead length is shown in Figure 10.5. As the effects of factor A and inter-

Table 10.12. Response table for the experiment

Trial no.	A	B	AB	C	D	AE	E	Crack length (mm)	
1	1	1	1	1	1	1	1	140	120
2	1	1	1	2	2	2	2	140	60
3	1	2	2	1	1	2	2	4	6
4	1	2	2	2	2	1	1	190	140
5	2	1	2	1	2	1	2	3	5
6	2	1	2	2	1	2	1	60	70
7	2	2	1	1	2	2	1	95	195
8	2	2	1	2	1	1	2	50	150

Table 10.13. Table of main and interaction effects for the experiment

Effects	Average crack length at level 1	Average crack length at level 2	Estimate of effect
A	100.00	78.50	−21.50
B	74.75	103.75	29.00
C	71.00	107.50	36.50
D	75.00	103.50	28.50
E	126.25	52.25	−74.00
A × B	118.75	59.75	−59.00
A × E	99.75	78.75	−21.00

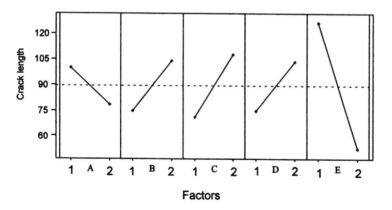

Figure 10.4. Main effects plot of the factors

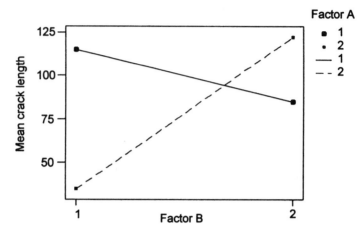

Figure 10.5.   Interaction plot between current and bead length

Table 10.14.   Pooled ANOVA table for the mean response

Source of variation	Degree of freedom	Sum of squares	Mean square	F-statistic	Percent contribution
B	1	3 364	3 364	1.84	2.32
C	1	5 329	5 329	2.91	5.29
D	1	3 249	3 249	1.77	2.14
E	1	21 904	21 904	11.96**	30.37
AB	1	13 924	13 924	7.60*	18.30
pooled error	10	18 317	1 831.70	—	41.58
Total	15	66 087	4 405.80	—	100

The critical F-values from tables are obtained as follows: $F_{0.05,1,10} = 4.96$, $F_{0.01,1,10} = 10.04$.
* —indicates that the factor or interaction effect is statistically significant at the 5% level of significance (i.e., 95% level of confidence).
** —indicates that the factor or interaction effect is statistically significant at both the 5% and 1% levels of significance (i.e., 95% and 99% levels of confidence).

action AE were relatively low in magnitude, they were pooled with the error.

The ANOVA table has shown that main effect E and interaction effect AB are statistically significant. Having constructed the ANOVA table, the optimal settings based on the mean response were determined as follows (see Table 10.13):

Table 10.15. SNR Table

Experimental run order	SNR values
1	−42.30
2	−40.64
3	−14.15
4	−44.45
5	−12.30
6	−36.28
7	−43.72
8	−40.97

$$A_2B_1C_1D_1E_2$$

Having performed the ANOVA on mean response, it was then decided to perform the SNR analysis. SNR was calculated corresponding to each experimental design point. As the objective of the experiment was to minimize the crack length, SNR for smaller-the-better quality characteristic was chosen (equation 9.11). Table 10.15 presents the SNR values based on this characteristic.

Having obtained the SNR values, the next step was to compute the average SNR value at each level of the factors and interactions and hence the effect of each factor or interaction on the SNR. Table 10.16 shows the average SNR and the effects of factors and interactions on the SNR. It shows that two main effects (C and E) and an interaction effect (A × B) have significant impact on the SNR. ANOVA was performed to see which of these effects are statistically significant. The results of the pooled ANOVA are presented in Table 10.17. The optimal factor settings based on the SNR were as follows:

$$A_2B_1C_1D_1E_2$$

The two analyses showed that there was no trade-off in factor levels. The final selection of factor settings for the process were:

- current: level 2 (135 amperes);
- bead length: level 1 (20 mm);
- electrode make: level 1 (type 1);
- V groove angle: level 1 (45 degrees);
- welding method: level 2 (method 2).

Table 10.16   Average SNR and estimates of effects

Main effects	Average SNR at level 1	Average SNR at level 2	Effect
A	−35.39	−33.32	2.09
B	−32.88	−35.82	−2.94
C	−28.12	−40.59	**−12.47**
D	−33.43	−35.28	−1.85
E	−41.68	−26.80	**14.88**
Interactions			
A × B	−41.91	−26.78	**15.12**
A × E	−35.01	−33.70	1.31

Table 10.17.   Pooled ANOVA Table for the SNR

Source of variation	Degrees of freedom	Sum of squares	Mean square	F-statistic	Percent contribution
C	1	310.90	310.90	34.47**	24.46
E	1	430.48	430.48	47.73**	34.15
A × B	1	456.74	456.74	50.64**	36.28
pooled error	4	36.08	9.02	—	5.11
Total	7	1234.21	176.32	—	100.00

From F-tables, $F_{0.05,1,4} = 7.71$ and $F_{0.01,1,4} = 21.20$.
** —indicates that the factor/interaction effect is significant at both 5% and 1% levels of significance (i.e., 95% and 99% levels of confidence).

Predicted SNR. The predicted SNR based on significant effects was given by:

$$\text{S}\hat{\text{N}}\text{R} = \overline{T} + (\overline{C}_1 - \overline{T}) + (\overline{E}_2 - \overline{T}) + [(\overline{A_2 B_1} - \overline{T}) - (\overline{A}_2 - \overline{T}) - (\overline{B}_1 - \overline{T})]$$

$$\text{S}\hat{\text{N}}\text{R} = \overline{C}_1 + \overline{E}_2 + \overline{A_2 B_1} - \overline{A}_2 - \overline{B}_1$$

$$= -28.12 + (-26.80) + (-27.02) + 33.32 + 32.88$$

$$= -15.74.$$

This is the predicted SNR based on the significant terms. The SNR based on the normal production settings was estimated to be −20.82. Therefore, gain in the SNR = 5.08 db.

CI for the SNR.   Here we construct a 95% CI for the SNR.

$$CI = \pm \sqrt{\frac{F_{(\alpha, 1, v_2)} \times MSE}{N_e}}$$

$F_{(0.05, 1, 4)}$=7.71; MSE = 9.02 (from Table 10.17).
  Effective number of replications = $N_e = 8/(1 + 3) = 2$.

$$\therefore CI = \pm \sqrt{\frac{7.71 \times 9.02}{2}} = 5.9.$$

Therefore, 95% CI for the predicted SNR = -15.74 ± 5.9 db.

Confirmatory run.   A confirmatory run based on the optimal factor settings was carried out (two replications) and crack lengths of 1.9 mm and 2.6 mm (SNR = -10.16) were observed. This has shown a significant improvement in terms of variability reduction in crack length and also reduction in mean crack length.

10.2.3 *Reducing variability in transformer inductance through Taguchi methods*

Nature of the problem.   High quality costs due to excessive variability in inductance from transformer to transformer.

Selection of the factors for the experiment.   The engineering team, consisting of people from quality, production and product design, have arrived at five control factors (A, B, C, D and E) which might have some impact on variability in inductance. Two noise factors (M and N) were also identified for the experiment. Due to confidentiality agreement between the authors and the organization, the names of the factors and the ranges of factor settings cannot be revealed in this study. Each factor was kept at two levels.

Selection of response.   Inductance.

Interactions of interest.   The interaction between the control factors B and C was of interest to the team.

Type of quality characteristic.   Nominal-the-best (type II).

Table 10.18.   Alias structure among the factors

Defining relationship: I = ABD = ACE = BCDE
Design resolution: III
Alias structure:

A = BD = CE	B = AD = CDE
C = AE = BDE	D = AB = BCE
E = AC = BCD	
BC = DE = ABE = ACD	
BE = CD = ABC = ADE	

Selection of OA for the experiment.   In this study, we have to deal with both control and noise arrays separately. As the experimenter wanted to study five control factors and one interaction effect, it was decided to select an L_8 OA (note that the degrees of freedom is equal to six). Similarly, an L_4 OA was selected for the noise array, as the team decided to incorporate two noise factors at two levels.

Design generators and aliasing structure.   The design generators for the OA design are given by: D = AB and E = AC. Therefore the design resolution of the design is III. This means that the main effects are confounded with two-factor interaction effects. The complete alias structure of the design is shown in Table 10.18.

Response table for the experiment.   The response table used for the experiment is illustrated in Table 10.19. Each of the eight trials in the control array was run against each of the four trials in the noise array. Hence the experiment consisted of a total of 32 experimental trials.

Statistical analysis and interpretation.   As the objective of the experiment was to reduce variability in inductance, the first step was to compute the SNR for each experimental trial (equation 9.12). The results are shown in Table 10.20.

Having calculated the SNR for each row, the next step was to develop a response table for the SNR. The SNR response table (Table 10.21) gives the average SNR for each factor and interaction at each level and also the effect. It shows that the most dominant main effect is factor A, followed by factor D, factor B, etc. Interaction between factors B and C also appeared to be very influential on variability in inductance. Figure 10.6 illustrates the main effects plot of the SNR.

Table 10.19.   Response Table for the experiment

Trial	A	B	D	C	E	BC	ABC	N M 1 1	1 1 2 1	2 1 1 2	1 2 2 2
1	1	1	1	1	1	1	1	8.4	7.21	9.12	8
2	1	1	1	2	2	2	2	9.34	7.82	10.42	7.55
3	1	2	2	1	1	2	2	10.11	8.2	7.92	9.52
4	1	2	2	2	2	1	1	7.9	8.75	8.15	10.22
5	2	1	2	1	2	1	2	10.2	10.44	9.92	11.02
6	2	1	2	2	1	2	1	8.85	9.52	9.04	7.88
7	2	2	1	1	2	2	1	7.54	8.71	9.73	8.15
8	2	2	1	2	1	1	2	8.22	9.9	9.14	8.88

Table 10.20.   SNR Table

Trial number	SNR
1	1.56
2	-2.58
3	-0.41
4	-0.34
5	6.84
6	2.90
7	0.63
8	3.15

Table 10.21.   Average SNR Response Table

	A	B	D	C	E	BC	ABC
Level 1	-0.44	2.18	0.69	2.16	1.80	2.80	1.19
Level 2	3.38	0.76	2.25	0.78	1.14	0.14	1.75
Effect	**3.82**	-1.42	1.56	-1.38	-0.66	**-2.66**	0.56

The interaction plot of BC is shown in Figure 10.7. For interaction effects, it is important to determine the average SNR for all possible combinations of factor levels. The average SNR for all factor level combinations of B and C is presented in Table 10.22.

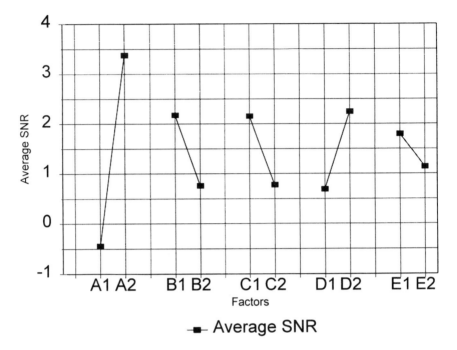

Figure 10.6.   Main effects plot of the SNR

Table 10.22.   Average SNR Table

Factor B	Factor C	Average SNR
B1	C1	4.20
B1	C2	0.16
B2	C1	0.11
B2	C2	1.40

Figure 10.7 shows that there is an interaction between factors B and C. In order to support this claim, ANOVA on the SNR was performed. The results of the pooled ANOVA are shown in Table 10.23. The ANOVA has shown that main effect A and interaction effect BC were statistically significant. Moreover, factor A was responsible for nearly 50% of the total

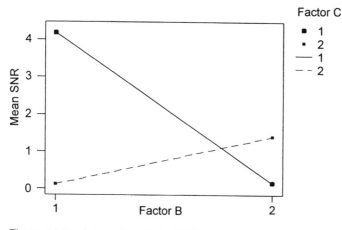

Figure 10.7.   Interaction Plot of BC

variation. Those effects with mean square (or variance) less than 4.5 were pooled to get a reasonable estimate of error variance.

Having performed the ANOVA for the SNR, the next step was to identify the optimum condition, which in this case is due to the factor settings that yield minimum variability in inductance.

Optimal factor settings based on the SNR.   The optimal factor settings are based on the SNR response table (Table 10.21). For main effect A, level 2 is the best (note that the higher the SNR, the lower the variability will be). For interaction BC, the best combination was level 1 for factor B and level 1 for factor C (Table 10.22). For insignificant factors, the determination of factor levels was based on economic grounds. The final selection of the factor settings was:

$$A_2B_1C_1D_2E_2.$$

Note that level 2 for factor E was chosen as it was more cost effective. This is the same as the factor settings defined in trial number 5 in Table 10.19.

Prediction of the SNR based on the optimal condition.   The prediction equation is generally based on the significant main or/and interaction effects. For the present case study, the prediction equation is based on the main effect 'A' and interaction effect BC, and is given by:

$$\hat{SNR} = \overline{T} + (\overline{A_2} - \overline{T}) + [(\overline{B_1 C_1} - \overline{T}) - (\overline{B_1} - \overline{T}) - (\overline{C_1} - \overline{T})]$$

where $\overline{T} = 1.47$.

$$\therefore \hat{SNR} = 1.47 + (3.38 - 1.47) + [(4.20 - 1.47) - (2.18 - 1.47) - (2.16 - 1.47)]$$
$$= 1.47 + 1.91 + 2.73 + 0.71 - 0.69$$
$$= 4.71 \text{ db.}$$

Confidence interval for the predicted SNR.   The confidence interval for the predicted SNR at the optimum condition is given by:

$$CI = \pm \sqrt{\frac{F_{(\alpha, 1, v_2)} \times MSE}{N_e}}.$$

Assume 95% CI: the tabled value of F with 5% significance level, 1 degree of freedom for the numerator and 4 degrees of freedom for the denominator (i.e. the error term) is 7.71. The error variance or mean square due to error term is 2.34 (Table 10.23). The effective number of replications is given by:

$$N_e = 8/(1 + 2) = 2.667.$$

Substituting the values into the above equation, we obtain:

Table 10.23.   Pooled ANOVA for the SNR

Source of variation	Degree of freedom	Sum of squares	Mean square	F-ratio	Percent contribution
A	1	29.18	29.18	12.47**	46.56
B	[1]	4.05	4.05	—	—
C	[1]	3.81	3.81	—	—
D	1	4.87	4.87	2.08	4.40
E	[1]	0.87	0.87	—	—
BC	1	14.24	14.24	6.09*	20.65
ABC	[1]	0.63	0.63	—	—
pooled error	4	9.36	2.34	—	—
Total	7	57.65	8.24	—	100

$F_{0.10,1,4} = 4.54$, $F_{0.05,1,4} = 7.71$ and $F_{0.01,1,4} = 21.20$.
** —factor or interaction effect is statistically significant at both the 90% and 95% confidence levels (or 10% and 5% significance levels).
* —factor or interaction effect is statistically significant at the 90% confidence level (i.e., 10% significance level).

$$CI = \pm \sqrt{\frac{7.71 \times 2.34}{2.667}}$$
$$= \pm 2.60 \text{ db.}$$

Therefore the predicted SNR at the optimum condition is 4.71 ± 2.60 db at the 95% CI.

Having determined the CI for the SNR, it is then suggested that a confirmatory run is performed. The idea is to verify whether the predicted SNR matches with the results from the confirmation run or experiment. Twenty samples were taken, based on the optimal factor settings, and the SNR, sample variance and sample standard deviation were calculated. The results are shown in Table 10.24.

In order to see the improvement in terms of variability reduction, it was decided to compare the standard conditions with the optimal condition. The average SNR (from a sample of 20) based on the standard production settings was estimated to be 2.15 db. The corresponding values of variance and standard deviation were 0.61 and 0.78 respectively. The selection of optimal factor settings has reduced the variability in transformer inductance by over 65%.

Calculation of cost savings

SNR based on the standard production conditions = 2.15 db.

SNR based on the optimal factor settings = 6.92 db.

The costs associated with scrap, rework, etc. due to excessive variability in inductance per monthly production was estimated to be $10000.

The quality costs based on optimal settings can be calculated using the following equation:

$$Costs_{optimal} = Costs_{existing} \times \left(\frac{1}{2}\right)^{\frac{SNR_{new} - SNR_{existing}}{3}}$$
$$= \$10\,000 \times (1/2)^{1.59}$$
$$= \$10\,000 \times (0.332)$$
$$= \$3320.$$

Table 10.24.   Results of the confirmation run

SNR	6.92
Sample variance	0.203
Sample standard deviation	0.45

Therefore, cost savings per month = $10 000 − $3320
$$= \$6680.$$

∴ Cost savings per annum ≈ $80 160.

10.2.4 Optimization of machine performance using Taguchi methods

This is an example of Taguchi's attribute analysis for optimizing machine parameters with the aim of maximizing the number of good parts. Review of the technical literature provided by the machine manufacturer has shown no systematic method of obtaining optimum performance. The company often relied on maintenance personnel to adjust the machine parameter settings to minimize scrap rate, rework, etc. which would lead to high quality costs. This case study shows how Taguchi's attribute analysis has assisted the organization in maximizing the number of good parts produced by the machine.

Nature of the problem. High scrap rate due to non-optimal machine settings.

Selection of the factors and their levels for the experiment. Three factors were of interest to the experimenter. As this was the first Taguchi experiment performed on the machine, it was decided to study all factors at two levels. Table 10.25 illustrates the list of factors and their levels.

Interactions of interest. The following interactions were of interest to the experimenter prior to carrying out the experiment: injection speed × mould temperature and mould temperature × mould pressure.

Choice of OA design. L_8 OA.

Table 10.25 List of factors for the experiment

Factors	Factor labels	Units	Levels 1	2
Injection speed	B	rev/sec.	4.5	6.5
Mould temperature	A	°C	160	180
Mould pressure	C	Kg/cm²	40	50

Coded design matrix. The coded design matrix for the experiment is shown in Table 10.26.

Response table. Twenty samples were produced at each of the eight combinations of factors. That is, a total of 160 samples were produced. The experimenter decided to count the number of good parts and treated the information as classified attribute data. The samples obtained from the experiment were qualitatively rated and grouped into three categories (or classes): good, fair and bad, where only the bad units were totally unacceptable to the customers. Table 10.27 presents the results. Having constructed the response table, the next step was to compute the average number of units at each level of each factor for each category. Since each factor has two levels, we will be comparing the results of 80 units at one level versus another 80 units at the second level. The difference between

Table 10.26. Coded design matrix for the experiment

Trial no.	A	B	AB	C	AC	BC	ABC
1	1	1	1	1	1	1	1
2	1	1	1	2	2	2	2
3	1	2	2	1	1	2	2
4	1	2	2	2	2	1	1
5	2	1	2	1	2	1	2
6	2	1	2	2	1	2	1
7	2	2	1	1	2	2	1
8	2	2	1	2	1	1	2

Table 10.27. Response Table for the experiment

Trial	A	B	AB	C	AC	BC	ABC	Good	Fair	Bad	Total
1	1	1	1	1	1	1	1	2	4	14	20
2	1	1	1	2	2	2	2	6	3	11	20
3	1	2	2	1	1	2	2	8	4	8	20
4	1	2	2	2	2	1	1	14	2	4	20
5	2	1	2	1	2	1	2	11	3	6	20
6	2	1	2	2	1	2	1	7	5	8	20
7	2	2	1	1	2	2	1	5	6	9	20
8	2	2	1	2	1	1	2	3	6	11	20

level 1 and level 2 for each category must be calculated. This gives the effect of each factor on each category.

Sample calculation. For factor A at level 1:

- Category, good: $A_1 = 2 + 6 + 8 + 14 = 30$.
- Category, fair: $A_1 = 4 + 3 + 4 + 2 = 13$.
- Category, bad: $A_1 = 14 + 11 + 8 + 4 = 37$.
- Total $= 30 + 13 + 37 = 80$.

Similarly, at level 2:

- Category, good: $A_2 = 11 + 7 + 5 + 3 = 26$.
- Category, fair: $A_2 = 3 + 5 + 6 + 6 = 20$.
- Category, bad: $A_2 = 6 + 8 + 9 + 11 = 34$.
- Total $= 26 + 20 + 34 = 80$.

Effect of factor A:

- Category, good: difference $= 30 - 26 = 4$.
- Category, fair: difference $= 20 - 13 = 7$.
- Category, bad: difference $= 37 - 34 = 3$.

By performing these calculations for each of the other factors and factor interactions, the classified attribute response table is generated (Table 10.28). The response block diagram based on all categories (sometimes called a response graph) is shown in Figure 10.8.

Table 10.28. Classified Attribute Response Table

Category		A	B	AB	C	AC	BC	ABC
Good	Level 1	30	26	16	26	20	30	28
	Level 2	26	30	40	30	36	26	28
	Effect	**4**	**4**	**24**	**4**	**16**	**4**	**0**
Fair	Level 1	13	15	19	17	19	15	17
	Level 2	20	18	14	16	14	18	16
	Effect	**7**	**3**	**5**	**1**	**5**	**3**	**1**
Bad	Level 1	37	39	45	37	41	35	35
	Level 2	34	32	26	34	30	36	36
	Effect	**3**	**7**	**19**	**3**	**11**	**1**	**1**

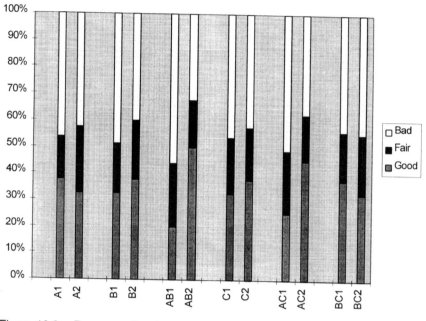

Figure 10.8.   Response Graph

As we need to maximize the number of good parts, the analysis was focused on that category. Table 10.28 has shown that none of the main effects is important. However, two interactions (AB and AC) were found to be very important. Having identified the most important effects, the next step was to determine the optimal factor level settings to maximize the number of good parts.

As the two interaction effects were important, it was decided to analyse them by obtaining the number of good parts at each level combination of AB and AC (Table 10.27). The results (Table 10.29) show that level 1 for factor A, level 2 for factor B and level 2 for factor C are the best combination. Therefore the optimal factor settings are:

$$A_1B_2C_2.$$

The average fraction acceptable for good category is obtained by:

$$\overline{T} = \frac{(2+6+8+\ldots+5+3)}{160}$$
$$= 0.35 \text{ or } 35\%.$$

Table 10.29.   Interaction table

Factor level combination	No. of good parts
A_1B_1	8
A_1B_2	22
A_2B_1	18
A_2B_2	8
A_1C_1	10
A_1C_2	20
A_2C_1	16
A_2C_2	10

The prediction equation becomes:

$$\hat{\mu} = \overline{T} + [(\overline{A_2 B_2} - \overline{T}) - (\overline{A_1} - \overline{T}) - (\overline{B_2} - \overline{T})]$$
$$+ [(\overline{A_1 C_2} - \overline{T}) - (\overline{A_1} - \overline{T}) - (\overline{C_2} - \overline{T})].$$

Convert the responses for the recommended levels into percentage occurrence:

$$\overline{A_1 B_2}(\%) = \frac{22}{40} = 55\%; \ \overline{A_1 C_2}(\%) = \frac{20}{40} = 50\%; \ \overline{A_1}(\%) = \frac{30}{80} = 37.5\%;$$

$$\overline{B_2}(\%) = \frac{30}{80} = 37.5\%;$$

$$\overline{C_2}(\%) = \frac{30}{80} = 37.5\%.$$

Substituting the values into the prediction equation, we get:

$$\hat{\mu} = \overline{T} + [(\overline{A_1 B_2} - \overline{T}) - (\overline{A_1} - \overline{T}) - (\overline{B_2} - \overline{T})]$$
$$+ [(\overline{A_1 C_2} - \overline{T}) - (\overline{A_1} - \overline{T}) - (\overline{C_2} - \overline{T})]$$
$$= 60\% \text{ (which is a meaningful result)}.$$

This shows that the predicted level of good parts is about 60%. Here we have not used omega transformation, as the transformation has better additivity in the neighbourhood of 0 or 1. It is important to note that the omega transformation transforms a range of 0 to 1 a range of $-\infty$ to $+\infty$.

The next step was to conduct a confirmation run and compare the actual results to the predicted results. If the results of the confirmation run or experiment are close to the predicted value, the optimal factor settings can then be established. On the other hand, if the results from the confirmation run are disappointing, further investigation may then be required.

For the present study, 50 samples were made using the optimal factor settings. The percentage of good parts was estimated as 64%, which is close to the predicted value. The cost of rework and scrap based on the existing machine parameter settings was estimated to be approximately $185,000 per annum. The estimated tangible annual savings when the process operated at the optimal settings was about $100,000. The case study has shown the importance of industrial designed experiments to the management. Moreover, top management have learned the lesson that trial and error experiments (which have been used by the engineers for several years within the organization) are inefficient, unreliable, time consuming and not cost-effective.

These case studies have demonstrated the importance and potential of carefully planned and designed experiments. In a rapidly changing business environment, the need for such an approach aimed at reducing process variability and thereby improving process performance is paramount, rather than the popular *ad hoc* approach using trial and error (or hit or miss) experimentation. Organizations can no longer afford to waste money by incurring high failure costs on the scale seen in the past. It has been the intention of the authors in producing this book to help in identifying and demonstrating good and sound practice so that organizations can not only survive into the new millennium but also become increasingly competitive.

APPENDICES

Appendix A
STANDARD ORTHOGONAL ARRAYS, INTERACTION TABLES AND LINEAR GRAPHS

Source: Taguchi, G. and Konishi, S. (1987) *Orthogonal Arrays and Linear Graphs*, ASI Press. Reproduced by permission of ASI, Michigan.

$L_4(2^3)$ Orthogonal Array

a) Coded design matrix:

Trial No.	Column		
	1	*2*	*3*
1	1	1	1
2	1	2	2
3	2	1	2
4	2	2	1

b) Interaction table:

		Column	
Column	1	2	3
1	[1]	3	2
2		[2]	1
3			[3]

c) Linear graph:

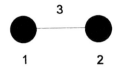

$L_8(2^7)$ Orthogonal Array

a) Coded design matrix:

			Column				
Trial no.	1	2	3	4	5	6	7
1	1	1	1	1	1	1	1
2	1	1	1	2	2	2	2
3	1	2	2	1	1	2	2
4	1	2	2	2	2	1	1
5	2	1	2	1	2	1	2
6	2	1	2	2	1	2	1
7	2	2	1	1	2	2	1
8	2	2	1	2	1	1	2

b) Interaction table:

	Column						
Column	1	2	3	4	5	6	7
1	[1]	3	2	5	4	7	6
2		[2]	1	6	7	4	5
3			[3]	7	6	5	4
4				[4]	1	2	3
5					[5]	3	2
6						[6]	1
7							[7]

c) Linear graphs:

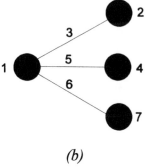

(a)

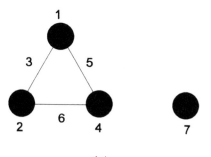

(b)

$L_{12}(2^{11})$ Orthogonal Array

Coded design matrix:

Trial no.	Column										
	1	2	3	4	5	6	7	8	9	10	11
1	1	1	1	1	1	1	1	1	1	1	1
2	1	1	1	1	1	2	2	2	2	2	2
3	1	1	2	2	2	1	1	1	2	2	2
4	1	2	1	2	2	1	2	2	1	1	2
5	1	2	2	1	2	2	1	2	1	2	1
6	1	2	2	2	1	2	2	1	2	1	1
7	2	1	2	2	1	1	2	2	1	2	1
8	2	1	2	1	2	2	2	1	1	1	2
9	2	1	1	2	2	2	1	2	2	1	1
10	2	2	2	1	1	1	1	2	2	1	2
11	2	2	1	2	1	2	1	1	1	2	2
12	2	2	1	1	2	1	2	1	2	2	1

$L_{16}(2^{15})$ Orthogonal Array

a) Coded design matrix:

Trial	Column														
	1	2	3	4	5	6	7	8	9	10	11	12	13	14	15
1	1	1	1	1	1	1	1	1	1	1	1	1	1	1	1
2	1	1	1	1	1	1	1	2	2	2	2	2	2	2	2
3	1	1	1	2	2	2	2	1	1	1	1	2	2	2	2
4	1	1	1	2	2	2	2	2	2	2	2	1	1	1	1
5	1	2	2	1	1	2	2	1	1	2	2	1	1	2	2
6	1	2	2	1	1	2	2	2	2	1	1	2	2	1	1
7	1	2	2	2	2	1	1	1	1	2	2	2	2	1	1
8	1	2	2	2	2	1	1	2	2	1	1	1	1	2	2
9	2	1	2	1	2	1	2	1	2	1	2	1	2	1	2
10	2	1	2	1	2	1	2	2	1	2	1	2	1	2	1
11	2	1	2	2	1	2	1	1	2	1	2	2	1	2	1
12	2	1	2	2	1	2	1	2	1	2	1	1	2	1	2
13	2	2	1	1	2	2	1	1	2	2	1	1	2	2	1
14	2	2	1	1	2	2	1	2	1	1	2	2	1	2	1
15	2	2	1	2	1	1	2	1	2	2	1	2	1	1	2
16	2	2	1	2	1	1	2	2	1	1	2	1	2	2	1

b) Interaction table:

Column										Column					
	1	2	3	4	5	6	7	8	9	10	11	12	13	14	15
1	[1]	3	2	5	4	7	6	9	8	11	10	13	12	15	14
2		[2]	1	6	7	4	5	10	11	8	9	14	15	12	13
3			[3]	7	6	5	4	11	10	9	8	15	14	13	12
4				[4]	1	2	3	12	13	14	15	8	9	10	11
5					[5]	3	2	13	12	15	14	9	8	11	10
6						[6]	1	14	15	12	13	10	11	8	9
7							[7]	15	14	13	12	11	10	9	8
8								[8]	1	2	3	4	5	6	7
9									[9]	3	2	5	4	7	6
10										[10]	1	6	7	4	5
11											[11]	7	6	5	4
12												[12]	1	2	3
13													[13]	3	2
14														[14]	1
15															[15]

c) Linear graphs:

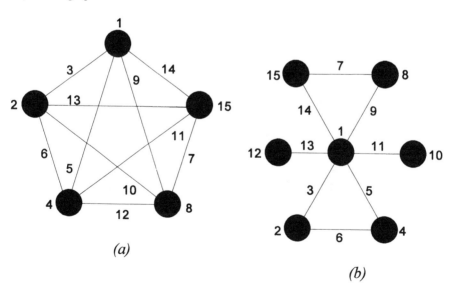

(a)

(b)

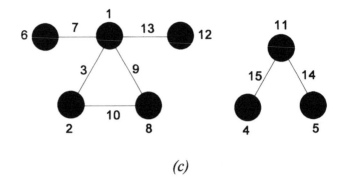

(c)

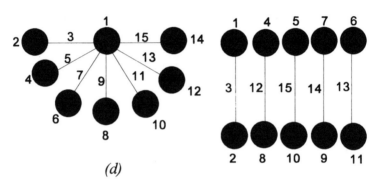

(d)

(e)

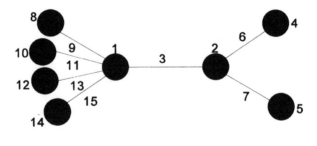

(f)

Appendix B
DESIGN GENERATORS FOR TWO-LEVEL OA OR FRACTIONAL FACTORIAL DESIGNS (NUMBER OF FACTORS VARY FROM 3 TO 7)

Source: *Minitab Reference Manual*, Minitab 11.0, Minitab Ltd. Reprinted by permission of Minitab Ltd., Westwood Business Park, Coventry, UK.

Number of trials	Number of factors				
	3	4	5	6	7
4	C = ±AB				
8		D = ±ABC	D = ±AB	D = ±AB	D = ±AB
			E = ±AC	E = ±AC	E = ±AC
			F = ±BC	F = ±BC	F = ±BC
					G = ±ABC
16			E = ±ABCD	E = ±ABC	E = ±ABC
				F = ±BCD	F = ±BCD
					G = ±ACD
32				F = ±ABCDE	F = ±ABCD
					G = ±ABDE

Appendix C
PERCENTAGE POINTS OF
THE F DISTRIBUTION

Source: Mitra, A. (1993) *Fundamentals of Quality Control and Improvement*, Macmillan Publishers, New York. Reproduced by permission of Macmillan Publishing Company, New York.

v_2	α	v_1								
		1	*2*	*3*	*4*	*5*	*6*	*7*	*8*	*9*
1	5%	161.4	199.5	215.7	224.6	230.2	234.0	236.8	238.4	240.5
	1%	4052.2	4999.5	5403.4	5624.6	5763.6	5859	5928.4	5981.1	6022.5
2	5%	18.51	19.00	19.16	19.25	19.30	19.33	19.35	19.37	19.38
	1%	98.50	99.00	99.17	99.25	99.30	99.33	99.36	99.37	99.39
3	5%	10.13	9.55	9.28	9.12	9.01	8.94	8.89	8.85	8.81
	1%	34.12	30.82	29.46	28.71	28.24	27.91	27.67	27.49	27.35
4	5%	7.71	6.94	6.59	6.39	6.26	6.16	6.09	6.04	6.00
	1%	21.20	18.00	16.69	15.98	15.52	15.21	14.98	14.80	14.66
5	5%	6.61	5.79	5.41	5.19	5.05	4.95	4.88	4.82	4.77
	1%	16.26	13.27	12.06	11.39	10.97	10.67	10.46	10.29	10.16
6	5%	5.99	5.14	4.76	4.53	4.39	4.28	4.21	4.15	4.10
	1%	13.75	10.92	9.78	9.15	8.75	8.47	8.26	8.10	7.98

v_2	α	1	2	3	4	5	6	7	8	9
7	5%	5.59	4.74	4.35	4.12	3.97	3.87	3.79	3.73	3.68
	1%	12.25	9.55	8.45	7.85	7.46	7.19	6.99	6.84	6.72
8	5%	5.32	4.46	4.07	3.84	3.69	3.58	3.50	3.44	3.39
	1%	11.26	8.65	7.59	7.01	6.63	6.37	6.18	6.03	5.91
9	5%	5.12	4.26	3.86	3.63	3.48	3.37	3.29	3.23	3.18
	1%	10.56	8.02	6.99	6.42	6.06	5.80	5.61	5.47	5.35
10	5%	4.96	4.10	3.71	3.48	3.33	3.22	3.14	3.07	3.02
	1%	10.04	7.56	6.55	5.99	5.64	5.39	5.20	5.06	4.94
11	5%	4.84	3.98	3.59	3.36	3.20	3.09	3.01	2.95	2.90
	1%	9.65	7.21	6.22	5.67	5.32	5.07	4.89	4.74	4.63
12	5%	4.75	3.89	3.49	3.26	3.11	3.00	2.91	2.85	2.80
	1%	9.33	6.93	5.95	5.41	5.06	4.82	4.64	4.50	4.39
13	5%	4.67	3.81	3.41	3.18	3.03	2.92	2.83	2.77	2.71
	1%	9.07	6.70	5.74	5.21	4.86	4.62	4.44	4.30	4.19
14	5%	4.60	3.74	3.34	3.11	2.96	2.85	2.76	2.70	2.65
	1%	8.86	6.51	5.56	5.04	4.69	4.46	4.28	4.14	4.03
15	5%	4.54	3.68	3.29	3.06	2.90	2.79	2.71	2.64	2.59
	1%	8.68	6.36	5.42	4.89	4.56	4.32	4.14	4.00	3.89
16	5%	4.49	3.63	3.24	3.01	2.85	2.74	2.66	2.59	2.54
	1%	8.53	6.23	5.29	4.77	4.44	4.20	4.03	3.89	3.78
17	5%	4.45	3.59	3.20	2.96	2.81	2.70	2.61	2.55	2.49
	1%	8.40	6.11	5.18	4.67	4.34	4.10	3.93	3.79	3.68
18	5%	4.41	3.55	3.16	2.93	2.77	2.66	2.58	2.51	2.46
	1%	8.29	6.01	5.09	4.58	4.25	4.01	3.84	3.71	3.60
19	5%	4.38	3.52	3.13	2.90	2.74	2.63	2.54	2.48	2.42
	1%	8.18	5.93	5.01	4.50	4.17	3.94	3.77	3.63	3.52
20	5%	4.35	3.49	3.10	2.87	2.71	2.60	2.51	2.45	2.39
	1%	8.10	5.85	4.94	4.43	4.10	3.87	3.70	3.56	3.46
21	5%	4.32	3.47	3.07	2.84	2.68	2.37	2.49	2.42	2.37
	1%	8.02	5.78	4.87	4.37	4.04	3.81	3.64	3.51	3.40
22	5%	4.30	3.44	3.05	2.82	2.66	2.55	2.46	2.40	2.34
	1%	7.95	5.72	4.82	4.31	3.99	3.76	3.59	3.45	3.35
23	5%	4.28	3.42	3.03	2.80	2.64	2.53	2.44	2.37	2.32
	1%	7.88	5.66	4.76	4.26	3.94	3.71	3.54	3.41	3.30
24	5%	4.26	3.40	3.01	2.78	2.62	2.51	2.42	2.36	2.30
	1%	7.82	5.61	4.72	4.22	3.90	3.67	3.50	3.36	3.26
25	5%	4.24	3.39	2.99	2.76	2.60	2.49	2.40	2.34	2.28
	1%	7.77	5.57	4.68	4.18	3.85	3.63	3.46	3.32	3.22
26	5%	4.23	3.37	2.98	2.74	2.59	2.47	2.39	2.32	2.27
	1%	7.72	5.53	4.64	4.14	3.82	3.59	3.42	3.29	3.18
27	5%	4.21	3.35	2.96	2.73	2.57	2.46	2.37	2.31	2.25
	1%	7.68	5.49	4.60	4.11	3.78	3.56	3.39	3.26	3.15
28	5%	4.20	3.34	2.95	2.71	2.56	2.45	2.36	2.29	2.24
	1%	7.64	5.45	4.57	4.07	3.75	3.53	3.36	3.23	3.12

V_2	α	1	2	3	4	5	6	7	8	9
						V_1				
29	5%	4.18	3.33	2.93	2.70	2.55	2.43	2.35	2.28	2.22
	1%	7.60	5.42	4.54	4.04	3.73	3.50	3.33	3.20	3.09
30	5%	4.17	3.32	2.92	2.69	2.53	2.42	2.33	2.27	2.21
	1%	7.56	5.39	4.51	4.02	3.70	3.47	3.30	3.17	3.07
40	5%	4.08	3.23	2.84	2.61	2.45	2.34	2.25	2.18	2.12
	1%	7.31	5.18	4.31	3.83	3.51	3.29	3.12	2.99	2.89
50	5%	4.03	3.18	2.79	2.56	2.40	2.29	2.20	2.13	2.07
	1%	7.17	5.06	4.20	3.72	3.41	3.19	3.02	2.89	2.78
60	5%	4.00	3.15	2.76	2.53	2.37	2.25	2.17	2.10	2.04
	1%	7.08	4.98	4.13	3.65	3.34	3.12	2.95	2.82	2.72
80	5%	3.96	3.11	2.72	2.49	2.33	2.21	2.13	2.06	2.00
	1%	6.96	4.88	4.04	3.56	3.26	3.04	2.87	2.74	2.64
90	5%	3.95	3.10	2.71	2.47	2.32	2.20	2.11	2.04	1.99
	1%	6.93	4.85	4.01	3.53	3.23	3.01	2.84	2.72	2.61
100	5%	3.94	3.09	2.70	2.46	2.31	2.19	2.10	2.03	1.97
	1%	6.90	4.82	3.98	3.51	3.21	2.99	2.82	2.69	2.59
∞	5%	3.84	3.00	2.61	2.37	2.21	2.10	2.01	1.94	1.88
	1%	6.64	4.61	3.77	3.32	3.02	2.80	2.64	2.51	2.41

v_1 = degrees of freedom for the numerator; v_2 = degrees of freedom for the denominator; α = significance level.

Glossary

Additivity. The term additivity in industrial experiments refers to the independence of factor effects. The effect of additive factors occurs in the same direction (i.e. they do not interact).

Aliasing. An effect is said to be aliased, if it cannot be separated from another effect. It is an alternative terminology for a confounded effect.

Analysis of variance (ANOVA). This is a powerful technique to identify how much each factor contributes to the deviation of the result from the mean, the amount of variation produced by a change in factor levels and the amount of deviation due to random error. ANOVA is used to determine the extent to which the factors contribute to variation and to test the statistical significance of factorial effects.

Attribute quality characteristic. Attribute quality characteristics are not continuous variables, but can be classified on a discretely graded scale. They are often based on subjective judgements such as none/some/severe or good/fair/bad.

Larger-the-better. This is a quality characteristic which gives improved performance as the value of the quality characteristic increases.

Brainstorming. An integral part of Taguchi's experimentation process, the goal of which is to create a list of factors and levels that need to be examined in an experiment.

Cause-and-effect diagrams. A powerful tool for examining effects or problems to find out the causes and to point out possible areas where data can be collected.

Classified attribute. The type of quality characteristic that is divided into discrete classes rather than being measured on a continuous scale.

Confidence interval. An interval within which a parameter of interest is said to lie with a specified level of confidence. The extreme values of a confidence interval are called confidence limits.

Confidence level. The probability of the truth of a statement that a parameter value will lie within two confidence limits.

Confirmation experiment. These are the trials that are conducted to verify the validity of the experimental results. Confirmation experiments are usually run with the optimal factor level settings. The results are then compared with the predicted optimal condition.

Confounding. This is a condition in which experimental information on factors cannot be separated. The information becomes confused or mixed with other information.

Contribution ratio. This is a measure of the contribution of variability due to a source (main effect or interaction effect) to the total variability of the experimental results.

Control factor. Control factors are those which can be selected and controlled by the design or manufacturing engineer. The experimenter has control of during all phases, i.e. experimental, production and operational phases.

Cost-reduction factor. A factor that has no significant effect on both the mean and variability. The selection of levels of such factors is based on costs, convenience in setting up, etc.

Defining relationship. This is a statement of one or more factor word equalities used to determine the aliasing structure in a fractional factorial design.

Degrees of freedom. The number of independent (fair) comparisons that can be made within a set of data. For a factor, the degrees of freedom is one less than the number of levels of that factor. The number of degrees of freedom for an interaction effect is the product of the degrees of freedom associated with each factor. The number of degrees of freedom for an orthogonal array is one less than the number of experimental trials.

Effect (of a factor).   The change in response produced by a change in the levels of a factor.

Experiment.   An experiment is a series of tests performed to discover an unknown effect or establish a hypothesis.

Experimental design.   A powerful statistical technique for studying and understanding the process or system behaviour and thereby providing a deep and spontaneous insight into the way the system or process works.

Factorial experiment.   An experiment designed to examine the effect of one or more factors in a process, each factor being held at different levels so that their effects on the process can be observed. If the experimental design considers all the possible factor level combinations under study, then the experiment is called a full-factorial experiment.

Factors.   These are variables which are believed to have an impact on the process or product performance. Examples of experimental factors include temperature, pressure, speed, material, operator, time, etc.

Fractional factorial experiment.   An experimental design that consists of a fraction of a full factorial experiment (i.e. a fraction of all factor level combinations).

F-ratio.   This is a measure of the variation contributed by a factor as compared with the error variance from which a level of confidence can be established.

F-table.   Provides a means for determining the significance of a factor to a specified level of confidence by comparing the calculated F-ratio to that from the F-distribution. If the calculated F-ratio is greater than the table value, the effect is then deemed to be significant.

Inner array.   This is used in parameter design to identify the combinations of control factors to be studied in a designed experiment. Also called a control array or design array.

Interaction.   Interaction is said to exist when the effect of one factor on the output depends on the level of another factor. In other words, an interaction occurs when two or more factors together have an effect on the output that is different from those of the factors individually.

Level of a factor. This is the factor setting. It is the specified value of a factor which can be used in an experiment. It may be measured qualitatively or quantitatively.

Linear graph. This is a pictorial representation devised by Dr Taguchi, used to facilitate the assignment of factors and interacting columns in orthogonal arrays. Factors are assigned to dots. An interaction is assigned to the line segment connecting dots.

Loss function. This is a method used for evaluating the quality of a product, especially when the performance of a product deviates from its target value. It is a quantitative means of evaluating quality in terms of cost.

Loss to society. Loss to society is different for manufacturers and consumers. Manufacturers: loss of market share of business, rework, scrap, reputation for company, down time, etc. Consumers: stress, inconvenience, dissatisfaction, time, etc.

Manufacturing tolerance. The assessment of the tolerance prior to shipping. The manufacturing tolerance is usually tighter than the consumer tolerance.

Mean square. This is the average squared deviation from the sample average calculated by the sum of squares divided by the number of degrees of freedom. In other words, it is an unbiased estimate of a population variability of a factor.

Mean square deviation. A measure of variability around the target value.

Mean square due to error. The variance considered as the error.

Measurement accuracy. This is the difference between the average result of a measurement with a particular instrument and the true value of the quantity being measured.

Measurement error. This is the difference between the actual and measured value of measured quantity.

Measurement precision. Measure of repeatability.

Noise. Noise is any source of uncontrolled variation which affects the quality characteristic or output of a product or process.

Noise factors. Any uncontrollable factors that cause product performance to vary.

Off-line quality control. Quality control activities applied before production, at the product-development stage, process-development stage or during installation or recommissioning of a process.

On-line quality control. Quality control activities which take place after production begins. These activities include manufacturing, quality assurance and customer support.

Orthogonal array. A matrix of numbers arranged in rows and columns. Each column represents a specific factor or condition that can be changed from experiment to experiment. Each row represents the state of the factors in a given experiment. The array is called orthogonal because the effects of the various factors can be separated from the effects of the other factors. Thus orthogonal array is a balanced matrix of factors and levels, such that the effect of any factor is not confounded with any other factor. In an orthogonal array for factors at two levels, for every pair of columns, every combination of factor levels appears the same number of times.

Optimization. This is the process of finding the best levels for each factor with the strongest effects according to the quality characteristic under consideration.

Outer array. A layout for the noise factors selected for experimentation. This is also called the noise array.

Parameter design. The second stage in Taguchi's off-line quality engineering system, where the best nominal settings of control factors are determined. It is used to identify the parameter settings of a process or product that will make the process or product robust (i.e. insensitive to noise).

Percent contribution. This yields the contribution of variability by a factor (in %) to the total variability. It is obtained by taking the ratio of the pure sum of squares due to a factor to the total sum of squares.

Pooling. This is a method of combining the insignificant factor or interaction effects during analysis which have little effect on the experimental results. The idea of pooling is to obtain adequate degrees of freedom for error.

Pure sum of squares. This is the true average effect of a factor which considers the degrees of freedom and the error mean square.

Quadratic loss function. This is an efficient and effective method for evaluating the quality loss in monetary units (say, dollars) due to deviation in the performance characteristic of a product from its target performance.

Quality. The loss imparted by a product to the society from the time the product is shipped to consumers.

Quality characteristic. This is the characteristic of a product or process that defines product or process quality. This is also called the functional characteristic. Examples are mileage of a vehicle, strength, tyre wear, porosity, etc.

Randomization. This is a technique applied to the order of conducting the experimental runs in order to minimize the effects of noise factors which are not included in the experiment. It also guards the experimental results against systematic biases.

Range of factor levels. The wider the range used in the experiment, the better is the chance of discovering the real effect of the variable on the quality characteristic. However, the wider the range, the less reasonable is the assumption of a linear effect for the variable.

Repeatability. This is a measure of variation in repeated measurements made by the same operator on a certain part using the same gauge.

Repetition. Repetition is preferred over replication when the cost of experimental set-up change is high. The trials are run in a systematic way rather than in a randomized manner. In repetition, the experiment may be repeated several times based on the same experimental set-up.

Reproducibility. This is a measure of variation in measurements obtained when several operators use one gauge to measure the identical characteristic on the same part.

Robust design.   A robust design is one where the output is insensitive to large variations in the input conditions.

Robustness.   This is a condition used to describe a product or process design that functions with limited variability in spite of uncontrollable sources of noise.

Signal factor.   A factor that can be used to adjust the output.

Signal-to-noise ratio (SNR).   A measure of the functional robustness of products and processes. The SNR originated in the communications field, in which it represented the power of the signal over the power of the noise. The formula used to compute the SNR depends on the type of quality characteristic being used. The higher the SNR, the smaller the quality loss will be.

Sum of squares.   The total of the squared differences from a set value, usually the mean.

System design.   The stage when the prototype for a product or process is developed. This prototype is used to define the initial settings of a product or process design characteristic.

Tolerance design.   Tolerance design is used when variability can no longer be reduced by parameter design. It involves investing money to reduce tolerances, e.g. use more expensive components, higher-grade material, etc.

Triangular table.   A table containing all the information needed to locate main effects and two-factor interactions. It is used for linear graph modification and the assignment of interactions.

Variance.   The mean square deviation (MSD) from the mean. The sum of squares divided by the degrees of freedom.

INDEX